Nanoscale Engineering in Agricultural Management

Editor

Ramesh Raliya

Research Scientist
Department of Energy, Environment and Chemical Engineering
Washington University in Saint Louis
St. Louis, Missouri, USA

CRC Press

Taylor & Francis Group
Boca Raton London New York

CRC Press is an imprint of the
Taylor & Francis Group, an **informa** business

A SCIENCE PUBLISHERS BOOK

CRC Press
Taylor & Francis Group
6000 Broken Sound Parkway NW, Suite 300
Boca Raton, FL 33487-2742

First issued in paperback 2020

© 2019 by Taylor & Francis Group, LLC
CRC Press is an imprint of Taylor & Francis Group, an Informa business

No claim to original U.S. Government works

ISBN-13: 978-1-138-56701-6 (hbk)
ISBN-13: 978-0-367-78001-2 (pbk)

Library of Congress Cataloging-in-Publication Data

Names: Raliya, Ramesh, editor.
Title: Nanoscale engineering in agricultural management / editor: Ramesh
    Raliya.
Description: Boca Raton, FL : CRC Press, Taylor & Francis Group, 2019. |
    Includes bibliographical references and index.
Identifiers: LCCN 2018061579 | ISBN 9781138567016 (hardback)
Subjects: LCSH: Plant biotechnology. | Nanobiotechnology. | Precision farming.
Classification: LCC TP248.27.P55 N36 2019 | DDC 630--dc23
LC record available at https://lccn.loc.gov/2018061579

Visit the Taylor & Francis Web site at
http://www.taylorandfrancis.com

and the CRC Press Web site at
http://www.crcpress.com

Preface

Agriculture is a vital piece of everyone's life, providing good food and health as well as the economic contributions. Today, agriculture is facing a number of challenges, mainly due to decreasing farm land, limited natural resources, climate change and growing demand for food as a result of an ever-increasing global population. Nanotechnology is an evolving science for the agricultural industry; it has the potential to bring the sustainable and precise solutions required for developing smart agriculture practices and address the challenges being faced by the ag-sector. Therefore, it is essential to have whole spectrum of knowledge of the new science. Hence, experts of the agricultural nanotechnology contributed their knowledge, covering plant growth, protection and management using engineering nanoscale materials, to the 11 chapter titles listed under the table of contents for the present book "Nanoscale Engineering in Agricultural Management". These chapters have been peer-reviewed and selected for publication based on the independent reviewers' report. The content of the book covers very specific, in-depth, and both fundamental and applied aspects from the latest ag-nanotechnology research. I congratulate all the contributing authors for their excellent work and its impact in the field of ag-nanotechnology. We hope that each chapter of the book will be very useful for the researchers, but also for the policymakers and other audiences from various backgrounds.

Ramesh Raliya

Contents

CHAPTER 1

Nanomaterials
Synthesis and Characterization

Ramesh Raliya,[1,*] *Vinod Saharan,*[2] *Kailash Choudhary,*[3,4]
Sudha Summarwar,[3] *Khushboo Gulecha,*[5]
Vikal Gupta[5] and *Prakash Mal Sain*[6]

1. What is Nanomaterial?

A material of nanoscale dimensions (10^{-9} meter) or one that is produced using nanotechnology is considered as nanomaterial. However, the definition of nanomaterial is constantly being revised as the technology progresses. The most widely accepted definition of nanomaterial is provided by the European Union— "A natural, incidental or manufactured material containing particles, in an unbound state or as an aggregate or as an agglomerate and where, for 50% or more of the particles in the number size distribution, one or more external dimensions is in the size range 1 nm–100 nm" (Commission 2018). This definition also provided exceptions for certain cases, such as environment health and safety concerns, where the number size distribution could be between 1 and 50%, and materials such as graphene or equivalents will be considered as nanomaterial if even one or more of their external dimensions are below 1 nm. Although, the United States Food and Drug Administration (US-FDA) defines nanomaterials as an engineered material or end product that has at least one dimension in the nanoscale range (approximately 1 nm to 100 nm); or an engineered material or end product which exhibits properties or phenomena, including physical or chemical properties or biological effects, that are

[1] Washington University in St. Louis, MO 63130, USA.
[2] Maharana Pratap University of Agriculture and Technology, Udaipur 313001, India.
[3] Academy of Translational Research, Jodhpur-342005 India.
[4] Institute of Advanced Studies, Pune-411008 India.
[5] Jai Narain Vyas University, Jodhpur-342001 India.
[6] Shri Vardhaman Jain Senior Secondary School, Jodhpur-342303 India.
* Corresponding author: rameshraliya@wustl.edu

attributable to its dimension(s), even if these dimensions fall outside the nanoscale range, up to one micrometer (US-FDA 2018). The definition of nanomaterials has evolved mainly on the basis of risk assessment or ingredient labelling. The main purpose of a firm definition of nanomaterials is to understand and explore the applications and implications which may affect the biosphere in both positive and deleterious ways.

2. How the Nano is Different from its Bulk Counterpart?

Nanotechnology involves the ability to engineer individual atoms and molecules of a nanostructure, thereby having more surface area to volume size ratio in comparison to the bulk counterpart of same material and the same mass. The dramatically increased surface area per mass of a nanomaterial produces a more reactive surface area than bulk particles, therefore, more and faster reactivity than a bulk particle of same mass.

3. Type of Nanomaterials—Natural, Incidental or Manufactured

As stated in the above definitions from the EU and the US-FDA, the nanomaterials could be natural, incidental or manufactured (Fig. 1). Natural nanoparticles include biological structures, i.e., Ribonucleic acids, monomers, intracellular organelles, pollens, viruses and several microbes; non-biological structures, such as particles from a volcano, sea-salt, water, etc. Besides this, nanomaterial could be classified as incidental or manufactured, based on their origin/source or type of manufacturing process. In a broader environmental terminology, some also call them anthropogenic particles. Their sources include synthesis/manufacturing, i.e., nanopowders of

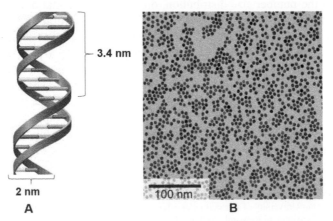

Fig. 1. Type of nanoparticles based on their origin/source or manufacturing process. (A) Sketch of Ribonucleic (DNA) structure with nanoscale dimensions, an example of natural nanostructure (B). Engineered gold nanoparticles manufactured by the solution of chemical reactions. The Fig. 1 B is obtained from (Raliya et al. 2017).

metal and metal oxides or by-product of various processes, such as combustion and wind. It should be noted that this chapter and the general perspective of the book is focused on engineered nanoparticles. Engineered nanoparticles can be further categorized based on their chemical composition, i.e., metal (e.g., gold, silver), metal oxide (e.g., Titanium or silicon dioxide), carbon (e.g., C-Dots, C60), polymer (e.g., chitosan nanoparticles), and structure, i.e., 2D (e.g., graphene, phosphine) and 3D nanostructures (crumpled graphene structure).

4. Good and Bad Nanoparticles

Nanoparticles or nanomaterials could be viewed as good or bad, depending on factors such as type, exposure concentration, risk profile, life cycle, stability in the system, surroundings and environment. For example, the incidental nanoparticles that originate as a result of combustion or volcanic events are harmful if inhaled in the excessive amount. These nanoparticles are known as bad particles because they can cause respiratory illnesses and damage the lung tissues of humans and animals. In contrast, engineered nanoparticles' properties are being used to detect diseases at early stages or diagnose the same. For example, nanoscale iron oxide showing super paramagnetic properties can image cancer cell in a body at early stage, help in effective diagnosis (Ferrari 2005). Similarly, certain size of nanoparticle can cross blood brain barrier being tested as drug carrier for rapid diagnosis, which is not possible in a non-invasive pathway of traditional approaches.

The toxicity or biocompatibility of nanoparticles determines its application. If the particles, generate reactive oxygen species (ROS), which are toxic to the cell survival could be used to kill cancer cell by combining with target delivery approaches. Biocompatible nanoparticles, in other words those particles which generate minimal ROS or non-reactive to biological system could be used as a therapeutic drug transporter. Further, nanoscale phenomenon can be tailored by engineering surface of the structure. The ROS generation can be tuned by controlling nanoparticles properties such as chemical composition, crystal phase and morphology in addition to specific surrounding environment including surface doping or coating. In a study, by Jiang et al. (Jiang et al. 2008, Jiang et al. 2009), established a method to investigate the dependence of the physicochemical properties of titanium dioxide (TiO_2) nanoparticles on their ROS generating capacity (Jiang et al. 2009). The size dependent ROS activity was established using different sizes (4–195 nm) but the same crystal phase. For a fixed total surface area, a sigmoidal shaped curve for ROS generation per unit surface area was observed as a function of particle size. The highest ROS activity per unit area was observed for 30 nm particles. The correlation between crystal phase and oxidant capacity was also established using TiO_2 nanoparticles. The ability of different crystal phases of TiO_2 nanoparticles to generate ROS was highest for amorphous, followed by anatase, and rutile crystal phase nanoparticles (Fig. 2).

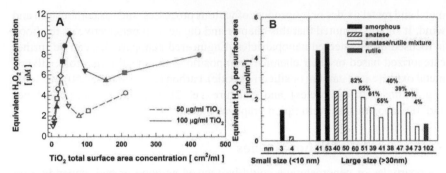

Fig. 2. The toxicity or biocompatibility of nanoparticles depends on their physicochemical properties. Generation of reactive oxygen species capability is a common criterion to determine the toxic profile of nanomaterial for a biological tissue. (A) Reactive oxygen species generation capacity depends on particle size and (B) Crystal phase affects ROS generation potential of nanoparticles having relatively the same size and chemical nature. The figure is obtained from (Jiang et al. 2008).

5. Manufacturing of Nanoparticles

The engineered nanoparticles are manufactured for the fundamental research and basic science applications. Today, commonly synthesized nanoparticles are being used commercially for biomedical, environmental, catalysis, agricultural, energy and information technology hardware. The most commercially used nanoparticles belong to three major classes—metal, metal oxide and carbon nanomaterial. In addition, organic nanostructures, such as organic polymers and mineral particles, are also used in biomedical engineering applications. In this section, we have discussed the synthesis methods for metal, metal oxide and carbon nanomaterials. There are two major manufacturing approaches commonly used for engineered nanoparticles: Bottom up fabrication, in which the starting materials are molecular precursors in the size range of angstrom and that can be allowed to nucleate and form the stable molecular cluster in the size range between 1 and 100 nanometers at least at one dimension. Whereas, top-down approach uses materials larger than 100 nanometers and eventually breaks them down to the scale of less than 100 nanometers.

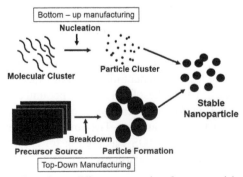

Fig. 3. Schematic illustration of two different approaches for nanoparticle synthesis—bottom-up fabrication (top) and top-down manufacturing.

Several methods have been developed in the past two decades to synthesize engineered nanoparticles with collectively or independently controlled characteristics. Table 1 summarizes the major synthesis routes for nanoparticles. The major factor determining the choice of a nanoparticle manufacturing process depends on the desired material properties and the cost. Chemical and aerosol, both bottom-up fabrications are being used to control nanoparticle properties. Although the manufacturing costs of aerosol-based, independently controlled particles are high, the process can be scaled up without compromising the quality of the material. Similarly, aerosol particles could cause toxicity to lungs and brain if inhaled and chemically synthesized nanoparticles may require additional downstream processing in order to become environmentally benign without inducing aggregation and agglomeration (Raliya et al. 2016). The biological and physical synthesis of nanoparticles has a limitation in controlling the properties of the materials, and hence may compromise the applications.

6. Characterization of the Particles

Nanomaterials' behavior in the provided environment, i.e., *in vivo* or *ex vivo* interactions, may be influenced by the physio-chemical properties, such as available surface area, functional groups, and morphology or crystal arrangements. Therefore, nanomaterial synthesis and independent characteristics play a vital role indetermining the nanomaterial's application and implications. Several versatile techniques have been developed to characterize nanomaterials with regards to their properties. Table 2 summarizes the available techniques routinely used to characterize, and their limitations.

7. Application of Nanomaterials

The ability to manipulate matter at nanoscale and understand its properties at the scale of 10^{-9} meter is leading to new technological development and revealing new properties of the existing material but at the reduced volume and increased surface area. Manufactured nanomaterials are being used and explored in virtually every sector, including agriculture, medicine, health, energy, transport, infrastructure, information technology hardware, environmental sustainability and space.

Among the myriad of nanotech application-based research reports, some examples are described below.

Food and Agriculture—A major problem of agriculture, fertilizer runoff that creates dead zones, can be potentially solved by the nanoscale fertilizer—showing enhanced uptake, minimize nutrient runoff and increase crop production (Raliya et al. 2016, Raliya et al. 2018), as well as agronomic fortification (Raliya and Tarafdar 2013, Raliya et al. 2015). Similarly, nanoscale chitosan combined with metal ions such as zinc or copper (Kumaraswamy et al. 2018) reduce the severity of the effect that pathogens have on the agriculturally important crops. In the future, nanotechnology

Table 1. Nanoparticle synthesis approaches.

Method	Synthesis Method			
	Physical	Biological	Chemical	Aerosol
Synthesis Type	Top-Down	Top-Down	Bottom-up	Bottom Up
Particle size distribution (PSD)	< 15–100 nm	< 100 nm	< 100 nm	< 100 nm
Polydispersity Index	High	Medium	Low	Low or high (depends on the method)
Advantages	• Rapid and scale up synthesis	• Environmentally benign • Surface coating with natural macro and micro biomolecules • Biocompatible	• Precisely control over morphology of metal nanoparticles	• Good control over particle size and shape • Synthesis of controlled nanocomposite • Monodispersity within a few percent • Passivation of the surface
Limitations	• Broad PSD and poly shape	• Rate of particle synthesis is low • Need natural resources • Broad PSD and poly shape	• Surface coating with harmful chemicals • Comparatively lesser bio-compatibility	• Large aggregates may be formed
Examples	• Ball Mills • Grinding	• Microbial synthesis • Plant extract mediated synthesis	• Sol-gel synthesis • Seed mediated approaches	• Furnace and Flame aerosol reactors • Room temperature atomization technology

Table 2. Techniques or instrumentation used to characterize nanomaterials.

Nanomaterials Properties	Microscopy Techniques	Spectroscopy Techniques	Notes
Morphology (size and shape)	Electron microscopy (TEM, SEM, STM, STEM) and AFM	UV-Vis Spectroscopy	Ultracentrifugation
Particle size distribution		DLS, SMPS	
Surface charge		DLS	In-flight DMA
Surface chemistry	AFM, CFM	FTIR, Raman Spectroscopy	
Surface area and porosity	Environmental – SEM, Cryo-SEM		BET
Crystal phase	HR-TEM	XRD	
Chemical composition	AEM, CFM	NMR, XPS, Auger AES, AAS, ICP-MS/OES, XRD, EBSD	

Abbreviations: STEM: Scanning Transmission Electron Microscope; TEM: Transmission Electron Microscopy; SEM: Scanning Electron Microscope; AFM: Atomic Force Microscope; STM: Scanning Tunneling Microscope; AEM: Analytical Electron Microscopy; CFM: Chemical Force Microscopy; XRD: X-ray Diffraction; NMR: Nuclear Magnetic Resonance; XPS: X-ray Photoelectron Spectroscopy; AES: Auger Electron Spectroscopy; AAS: Atomic Absorption Spectroscopy; ICP-MS/OES: Inductively Coupled Plasma Mass/Optical Emission Spectrometry; EBSD: Electron Backscatter Diffraction; DMA: Differential Mobility Analyzer; SMPS: Scanning Mobility Particle Sizer; BET: Brunauer–Emmett–Teller; FTIR: Fourier Transform Infra-red

may bring a concept of calorie food instead of biomass food to fulfill the food demand of over increasing global population. This can be realized by enhancing the nutritional value of the food products. Improved agronomic fortification will address malnutrition, a hidden hunger challenge, but also reduce the requirement of high volume of food; for example, a protein bar of a few grams can provide an equal amount of protein as received from few hundreds gram of other foods.

Environment—It has been explored that nanoscale multidimensional structures of graphene are used to understand nanoparticles' interaction with environment and purify water (Jiang et al. 2016, Jiang et al. 2017). Further, graphene combined with wood or bacterial cellulose was developed to desalinate the water in the presence of sun light (Jiang et al. 2016, Liu et al. 2017).

Medicine—Nanoscale materials are being used for drug delivery to deep tissues, such as the brain. Raliya et al. (Raliya et al. 2017) and Ye et al. (Ye et al. 2018) developed a non-invasive and target drug delivery technique by using gold nanoparticles combined with aerosol and focused ultrasound technology. Similarly, nanoscale materials are being used to control the cancer tumor growth by regulating the pH of microenvironments (Raliya et al. 2016, Som et al. 2016), or generating reactive oxygen species in the presence of Cerenkov radiations (Kotagiri et al. 2015). Furthermore, nanostructures of soft materials are being tested for the targeted drug carrier, enhanced blood circulation, real time imaging and non-invasive detection of the disease signals.

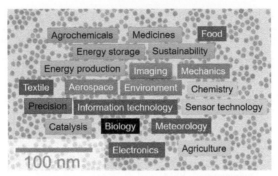

Fig. 4. Interdisciplinary applications of nanomaterials.

8. Policy and Regulations Related to Nanomaterial

In order to sustainable use and expand the potential application of new technology, including nanotechnology, it is important to consider both the potential benefits and the unintended or potential risks to human health and the environment that might accompany the development and use of the technology. Engineered nanomaterials are being used in wide range of consumer products, whereas natural or un-intended nanoscale materials could contaminate such consumer products. The prominent regulations to deal with the potential risk of nanoscale materials are laid down by the United States Environmental Protection Agency (US-EPA) and the Food and Drug Administration (US-FDA) or the Health and Consumer Protection Directorate of the European Union (EU) Commission. So far, these agencies and similar ones across the globe are limited to determining the definition of nanomaterials, and releasing advisory reports regarding safe use of nanomaterials with the aim of minimalizing human and environmental exposure to nanomaterials. The major limitations, keeping these regulations inconsistent over time and impeding a tangible conclusion, are the lack of data sets, reproducibility, and changing materials behavior based on physio-chemical properties. Therefore, it is important to understand the endpoints for the manufactured nanomaterials. As described in the Fig. 5, nanomaterials could end-up at multiple sites, depending on the type, use and application. A clear regulation in the harmony of global regulatory agencies will enable the responsible development of nanoscale-based products with new and beneficial properties. The regulatory science needs to develop a strategic plan of action for increasing data collection and analyses for the developing framework of the universal physico-chemical characterization, nanomaterial risk identification, assessment, communication and mitigation. It can be done by: (1) Developing improved methods and tools to detect and measure the physical structure, chemical properties, and safety of nanomaterials. (2) Developing and evaluating *in vitro* and *in vivo* assays and models to assess safety and/or efficacy of nanomaterials. (3) Incorporating the relevant risk characterization information, hazard identification, exposure science, and risk modeling and methods into the safety evaluation of nanomaterials.

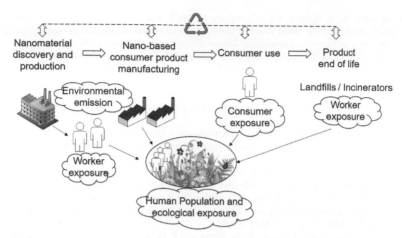

Fig. 5. Critical check points of a nanomaterial's life cycle, required for the environmental health and safety evaluation.

(4) Enhancing the state of knowledge and scientific evidence to support potential development of generalized class-based approaches to risk assessment of the products containing nanomaterials. (5) Integrating and standardizing risk communication within the risk management framework. (6) Developing methods for nanomaterial-based risk mitigation. The use of data science and artificial intelligence could be useful in assessing the available data and developing a reliable approach that not only promotes nanotechnology research, but also helps us use nanotechnology in a safe and sustainable manner.

References

Commission, E. 2018. Definition of a Nanomaterial. Retrieved October 28, 2018.

Ferrari, M. 2005. Cancer nanotechnology: Opportunities and challenges. Nature Reviews Cancer 5(3): 161.

Jiang, J., G. Oberdörster and P. Biswas. 2009. Characterization of size, surface charge, and agglomeration state of nanoparticle dispersions for toxicological studies. Journal of Nanoparticle Research 11(1): 77–89.

Jiang, J., G. Oberdörster, A. Elder, R. Gelein, P. Mercer and P. Biswas. 2008. Does nanoparticle activity depend upon size and crystal phase? Nanotoxicology 2(1): 33–42.

Jiang, Q., L. Tian, K.K. Liu, S. Tadepalli, R. Raliya, P. Biswas, R.R. Naik and S. Singamaneni. 2016. Foams: Bilayered biofoam for highly efficient solar steam generation (adv. Mater. 42/2016). Advanced Materials 28(42): 9234–9234.

Jiang, Y., R. Raliya, J.D. Fortner and P. Biswas. 2016. Graphene oxides in water: Correlating morphology and surface chemistry with aggregation behavior. Environmental Science & Technology 50(13): 6964–6973.

Jiang, Y., R. Raliya, P. Liao, P. Biswas and J.D. Fortner. 2017. Graphene oxides in water: Assessing stability as a function of material and natural organic matter properties. Environmental Science: Nano 4(7): 1484–1493.

Kotagiri, N., G.P. Sudlow, W.J. Akers and S. Achilefu. 2015. Breaking the depth dependency of phototherapy with cerenkov radiation and low-radiance-responsive nanophotosensitizers. Nature Nanotechnology 10(4): 370.

Kumaraswamy, R., S. Kumari, R.C. Choudhary, A. Pal, R. Raliya, P. Biswas and V. Saharan. 2018. Engineered chitosan based nanomaterials: Bioactivity, mechanisms and perspectives in plant protection and growth. International Journal of Biological Macromolecules.

Liu, K.-K., Q. Jiang, S. Tadepalli, R. Raliya, P. Biswas, R.R. Naik and S. Singamaneni. 2017. Wood–graphene oxide composite for highly efficient solar steam generation and desalination. ACS Applied Materials & Interfaces 9(8): 7675–7681.

Raliya, R. and J.C. Tarafdar. 2013. Zno nanoparticle biosynthesis and its effect on phosphorous-mobilizing enzyme secretion and gum contents in clusterbean (cyamopsis tetragonoloba l). Agricultural Research 2(1): 48–57.

Raliya, R., A. Som, N. Shetty, N. Reed, S. Achilefu and P. Biswas. 2016. Nano-antacids enhance ph neutralization beyond their bulk counterparts: Synthesis and characterization. RSC advances 6(59): 54331–54335.

Raliya, R., D. Saha, T.S. Chadha, B. Raman and P. Biswas. 2017. Non-invasive aerosol delivery and transport of gold nanoparticles to the brain. Scientific Reports 7: 44718.

Raliya, R., J.C. Tarafdar and P. Biswas. 2016. Enhancing the mobilization of native phosphorus in the mung bean rhizosphere using zno nanoparticles synthesized by soil fungi. Journal of Agricultural and Food Chemistry 64(16): 3111–3118.

Raliya, R., R. Nair, S. Chavalmane, W.-N. Wang and P. Biswas. 2015. Mechanistic evaluation of translocation and physiological impact of titanium dioxide and zinc oxide nanoparticles on the tomato (solanum lycopersicum l.) plant. Metallomics 7(12): 1584–1594.

Raliya, R., T. Singh Chadha, K. Haddad and P. Biswas. 2016. Perspective on nanoparticle technology for biomedical use. Current Pharmaceutical Design 22(17): 2481–2490.

Raliya, R., V. Saharan, C. Dimkpa and P. Biswas. 2018. Nanofertilizer for precision and sustainable agriculture: Current state and future perspectives. Journal of Agricultural and Food Chemistry 66: 6487–6503.

Som, A., R. Raliya, L. Tian, W. Akers, J.E. Ippolito, S. Singamaneni, P. Biswas and S. Achilefu. 2016. Monodispersed calcium carbonate nanoparticles modulate local ph and inhibit tumor growth *in vivo*. Nanoscale 8(25): 12639–12647.

US-FDA. 2018. FDA's approach to regulation of nanotechnology products. Retrieved October 28, 2018.

Ye, D., X. Zhang, Y. Yue, R. Raliya, P. Biswas, S. Taylor, Y.-C. Tai, J.B. Rubin, Y. Liu and H. Chen. 2018. Focused ultrasound combined with microbubble-mediated intranasal delivery of gold nanoclusters to the brain. Journal of Controlled Release 286: 145–153.

CHAPTER 2

Tools and Techniques for Characterizing Nanomaterial Internalization/Uptake in Plants and its Importance in Agricultural Applications

Remya Nair

1. Introduction

Plants play an important role in the ecosystem by holding a crucial interface between the humans and environment. Increased use of nanomaterials has created great concerns about the fate of nanoparticles, and direct interaction of nanoparticles with plants results in their internalization, which has a potential impact on crop growth and food safety. Plant nanotechnology utilizes nanomaterials to improve plant growth and to augment plant functions (Wu et al. 2017). Hence, it is high time to understand the fate of internalized nanomaterials in plant systems using several qualitative and quantitative analysis techniques. Research in nanotechnological applications in plant biology will benefit greatly from the developments in novel analytical methodologies which result in *in situ* analysis of nanoparticles in plants with high lateral resolution and low detection limit. Several microscopic and spectroscopic techniques can be used to determine the size, size distribution and concentration of internalized nanomaterials in plant tissues (Wang et al. 2016). The imaging techniques include the standard light microscopy techniques and

Nano Research Facility and Jens Lab, Department of Energy, Environment and Chemical Engineering, One Brookings Drive, Washington University in St. Louis, MO-63105, USA.
Email: rnair@wustl.edu, remya.d.nair@gmail.com

electron microscopy methods, such as Transmission Electron Microscopy (TEM) and Scanning Electron Microscopy (SEM), as well as several X-ray and spectral methods. Recent developments in electron microscopy with high lateral resolution help to detect individual nanoparticles in plant tissues, however the major limitation lies in the exact localization of particles in the tissues. Inductively Coupled Plasma (ICP) spectroscopy methods provide an exceptional detection limit of a wide range of elements in plant tissues. SP-ICP-MS (Single Particle Inductively Coupled Plasma Mass Spectrometry) is a very novel method which can be utilized for the simultaneous determination of size, size distribution and particle concentration in plant tissues with special enzymatic digestion method. Synchrotron-based X-ray spectroscopy methods can be successfully utilized for *in situ* assessment of chemical speciation in plant tissues (Wang et al. 2015). This chapter reviews the tools and techniques used to study nanoparticle interaction, uptake and distribution in plant system and its importance in agriculture.

2. Microscopy Tools for Detecting the Internalization of Nanomaterials

2.1 Light Microscopy

Light microscopy provides a good visualization of particles ranging in size from sub-micrometer to micrometer. It has several advantages over electron microscopy. The most important advantage is the ability to observe particles in live specimens without any treatment. Even though nanoparticles have size within the nanometer range, which is far below the resolution limit of a regular light or optical microscope, their presence in tissues as aggregates can be visualized with optical microscope facilities, such as Differential Interface Contrast (DIC) microscopy, confocal microscopy, fluorescence microscopy, etc. There were reports that had used light microscopy to image hand-cut sections of stem and petiole of treated plants (González-Melendi et al. 2007). Observance under phase contrast and epifluorescence at blue-violet excitation wavelength was reported for paraffin embedded samples sectioned with rotary microtome, and the nanoparticles were detected by differential interference contrast or reflection. Vibratome sectioned samples were imaged using a confocal laser scanning microscope, in which the samples were excited with Ar laser line of 488 nm and the reflected signal was collected in the rage of 480–500 nm. Corredor et al. also reported the use of light photo microscope under phase contrast, bright field and dark field to confirm the nanoparticle penetration and transport to various locations from the point of application in pumpkin plants. Phase contrast images from different regions confirmed the presence of nanoparticle aggregates (Corredor et al. 2009). Fluorescence microscopy was used to understand the uptake efficiency of quantum dots (QDs) by *Arabidopsis thaliana* root tissues. The microscopy images revealed that the majority of the QDs adsorbed on the zone of cellular maturation where lots of root hairs were present (Navarro et al. 2012). The uptake of QDs by rice (*Oryza sativa*) seedlings and magnetic iron oxide nanoparticles by corn plants were also confirmed with the help of fluorescent microscopy images (Nair et al. 2011, Li et al. 2016).

Microscopic evidence of the uptake and translocation of hematite and ferrihydrite nanoparticles in maize plants was reported, in which 3D microscopic techniques with confocal laser scanning microscope were used to confirm the translocation of engineered magnetic nanoparticles in maize plant tissues. Nanoparticle aggregates were observed in different plant regions, such as endodermis, xylem, phloem vessels and cell walls of xylem vessels (Pariona et al. 2017). Confocal laser scanning microscope was also used to record images confirming the uptake of water soluble upconversion nanoparticles by different plant species through roots to plant stalk and leaves. Microscopic studies confirmed that nanoparticles could enter the stele of plant roots, promoting long distance transport and *in vivo* distribution into the stem and leaves (Hischemöller et al. 2009). Lin and Xing used light microscopy to study the cell division states of root tips of ryegrass upon treatment with ZnO nanoparticles and Zn^{2+} ions (Lin and Xing 2008).

Confocal microscopy was also used to confirm the ability of fluorescein isothiocyanate (FITC) labelled single-walled carbon nanotubes (SWCNTs) to traverse across the cell wall and cell membrane of tobacco Bright Yellow (BY-2) cells in a temperature and time-dependent manner. This study opened a venues to utilize SWCNTs as agents to transport payloads effectively to different plant cell organelles (Liu et al. 2009). Confocal fluorescence imaging also confirmed the uptake of FITC-stained cerium dioxide (CeO_2)NP uptake by roots of corn (*Zea mays*) plants and nanoparticle aggregates were found in cell walls of epidermis and cortex (Zhao et al. 2012). Lin et al. studied the uptake, translocation and transmission of carbon nanomaterials in rice plants (Lin et al. 2009). Tissues at different developmental stages were sectioned and imaged using bright field microscope. Bright field images showed a greater presence of black aggregates of carbon nanomaterials in seeds and roots, and a lower amount in stem and leaves, which helped to track the transport. Detection and imaging of nanomaterials such as multiwalled carbon nanotubes (MWCNTs), titanium dioxide (TiO_2) and CeO_2 in living wheat (*Triticum aestivum*) tissues was carried out using two-photon excitation microscopy (TPEM) (Wild and Jones 2009). The nanosolutions were analyzed for optimum auto fluorescence excitation and emission between 690 nm and 800 nm using TPEM. The emission range for studied nanomaterials was detected between 360 and 660 nm and unstained aggregates of nanoparticles on wheat roots were visualized using TPEM and auto fluorescence. Uptake studies were performed with MWCNTs and 3D images of the penetration of MWCNTs and its interaction with phenanthrene, a model polycyclic aromatic compound, was visualized. This was one of the significant studies that tracked the chemical-nanomaterial interaction in living tissues. TPEM, along with auto fluorescence, provides a non-intrusive tool for the *in vivo* visualization of fate, interaction and behavior of nanomaterials in plant and animal tissues which could enhance several future drug delivery applications in animal and plant system. The uptake of silver nanoparticles (AgNPs) through roots was reported in *Arabidopsis thaliana* on treatment with 40 nm AgNPs. Imaging was done using a confocal/multiphoton microscope and the nanoparticles were observed in columella of plant roots (Ma et al. 2010). In another study by the same group with *Arabidopsis thaliana*, confocal laser scanning microscopy imaging of plant roots confirmed the uptake of 40 nm AgNPs by roots. Nanoparticles were observed in different root tissues which had led to more studies on transport of nanoparticles between cells within the root (Geisler-Lee et al. 2012). A large number of studies showed bright field images of different plant parts to confirm the uptake and translocation of nanomaterials by the presence of nanoparticle aggregates (Yan et al. 2013).

2.2 Atomic Force Microscope [AFM]

The invention of AFM helps to obtain real space images in the physiological aqueous environment of biological and biochemical structures with a resolution which is better than the diffraction limit of conventional optical microscopy (Chang et al. 2012). In English Ivy (*Hedera helix*), AFM studies demonstrated that nanoparticles were associated with the adhesive secretion from root hairs of adventitious roots, forming a natural nanocomposite. Analysis of AFM images in three-dimensional and phase mode confirmed that the nanoparticles associated with adhesive secretion were very uniform in topography and phase shifts. The importance of studying plant attachment systems to be applied for bio-inspired engineering was demonstrated in this research (Lenaghan and Zhang 2012). Abraham et al. used AFM to study the sorption of silver nanoparticles (AgNPs) to environmental and model surfaces. Morphological and nanomechanical parameters of the interaction of nanoparticles on the surface of *Ficusbenjamina* leaf disks were assessed with AFM. This study provided insights into the fate of engineered nanoparticles in environmental systems, which is controlled by changes in colloidal stability and their contact with various environmental surfaces (Abraham et al. 2013). Hence, AFM can be coupled with other techniques, such as electron microscopy, in order to quantify the size, size distribution and effective interaction of nanoparticles with different environmental surfaces.

2.3 Electron Microscopy

2.3.1 Transmission Electron Microscopy (TEM) and Scanning Electron Microscopy (SEM)

Ma et al. analyzed the root samples of *Arabidopsis thaliana* plants exposed to 20 nm AgNPs using TEM, three weeks after exposure. TEM images showed the accumulation of AgNPs at plasmodesmata. This confirmed that AgNPs were mostly taken to the intercellular spaces and transported inside the plant cells through plasmodesmata (Ma et al. 2010). Detection and quantification of carbon-based nanomaterials were also done with TEM and SEM. In this study, SEM images showed the adsorption of MWCNTs to the plant root surfaces of the treated plants due to their high affinity for the epidermis. TEM images of second generation plants' (raised from the seeds of first generation carbon nanomaterial treated plants) leaves showed the presence of C_{70} in leaf cells of two week old rice plants and, hence, investigated the generational transmission of carbon nanomaterials (Lin et al. 2009). Pot experiment was conducted with two types of TiO_2 NPs (NNM and P_{25}) on red clover plants in order to study the transport and uptake efficiency when grown in natural agricultural soil. The results were confirmed with the TEM micrographs which had showed the presence of NNM TiO_2 NPs on the outside of the root and P_{25} TiO_2 on the inside of the root and also transported to a different location (Gogos et al. 2016). The uptake of single-walled carbon nanohorns by tomato seeds/roots and tobacco cells was confirmed by TEM imaging (Lahiani et al. 2015). The confirmation of the uptake studies proposed the use of nanohorns as plant growth regulators due to their positive effects on the germination and further growth of several plant species. The uptake and distribution of MWCNTs by wheat and rapeseed (*Brassica napus*) was confirmed with the presence of nanotubes in the leaf ultrastructure by TEM imaging (Larue et al. 2012). TEM images of leaf tissues also confirmed the uptake of CeO_2 NPs by *Arabidopsis thaliana* plants grown in nanoparticle amended medium (Yang et al. 2017).

Electron microscopy imaging using high resolution TEM (HR-TEM) and Field emission SEM (FE-SEM) was used to confirm the accumulation of gold nanoparticles (AuNPs) in the rhizhodermis of barley (*Hordeum vulgare*) seedlings grown in hydroponics with 5 nm AuNPs (Milewska-Hendel et al. 2017). AuNPs were found to be accumulated around root cap cells and surface of the outer periclinal cell wall of the rhizhodermis and nanoparticles were not found in the any of the cell organelles irrespective of the tissue type. These findings promoted studies leading to answer several questions on more accumulation of AuNPs on root surface and non-translocation of AuNPs from the environment to inside of roots. The effects of AgNPs on the symbiosis of faba bean (*Vicia faba*) with *Rhizobium leguminosarum* bv. *Viciae* growth and *Glomus aggregatum* activity in the soil was investigated (Abd-Alla et al. 2016). In this study TEM and SEM were used in order to understand the retarded nodulation process and reduced mycorrhizal colonization and responsiveness. SEM images showed the presence of healthy bacteroids on untreated root nodules, whereas the root nodules of treated plants showed infected cells with deteriorating bacteroids. TEM images of AgNP-treated root nodules in the fixation zone showed a lower density of deformed bacteroids with occluded intercellular space and large central vacuole. It also showed the invagination of a vesicle-like structure towards the vacuole. The microscopic studies helped to interpret the mechanism of toxic action of AgNPs which had led to early senescence of root nodules. A similar study with nano-TiO_2 and nano-ZnO was earlier reported by Fan et al. and Huang et al., respectively, in which TEM images helped to understand the abnormalities in Rhizobium-legume symbiosis with TiO_2 treatment (Fan et al. 2014, Huang et al. 2014). Siani et al. investigated the influence of Arbuscular Mycorrhizal Fungus (AMF) on the growth and development of Fenugreek (*Trigonella foenumgraecum*) on treatment with various concentrations of ZnO nanoparticles and, in this study, TEM analysis was done to understand the cellular and subcellular damage in leaves due to nanoparticle localization upon exposure to high concentration of ZnO NPs (Siani et al. 2017). TEM images of leaf ultrastructure showed particle localization within the mesophyll cells with membrane damage. The adverse effects of MWCNTs on the leaf and root tissues of spinach (*Amaranthus tricolor* L.) were analyzed using SEM and TEM (Begum and Fugetsu 2012). The morphological evidence of broken midrib, epidermis swelling, irregularly shaped cells, and stomatal closure were observed with SEM analysis. Localization of CNTs inside the root and leaves was confirmed with TEM. Damaged membrane and degradation of granum and thylakoid were observed in TEM images of treated leaves. The uptake of plant synthesized AuNPs by rice seedlings was also confirmed by TEM imaging of root tissues, in which the nanoparticles were observed in membrane-bound compartments, such as the vacuole, which mitigated the severity of its toxic effects on plant tissues (Ndeh et al. 2017). TEM micrographs of the leaves of barley plants grown in CeO_2 and TiO_2 nanoparticle amended soil revealed the presence of nanoparticles in parenchyma leaf tissues, in chloroplast stroma and in the vacuoles. It was noticed that the chloroplasts still maintained its integrity after nanoparticle treatment (Marchiol et al. 2016). However, it was observed that the vascular parenchymal cells had detached plasma membrane, disorganized mitochondria, swollen cristae and lobe-shaped nucleus with condensed chromatin. After the verification of the translocation of Ce and Ti from roots to leaves of plants, Energy Dispersive X-ray spectroscopy (EDX) analysis was carried out in the selected region of leaf tissue in order to confirm the presence of Ce and Ti. Interestingly, Ce was not detected in leaf tissue as crystalline nanoclusters, and authors explained the possibility of nano CeO_2-induced massive formation of amorphous clusters of light elements rather than crystalline nano-CeO_2 clusters. TEM images also confirmed the uptake and translocation of magnetic iron oxide nanoparticles in corn (*Zea mays* L.) plants. Microscopic images revealed that magnetic nanoparticles could enter through plant roots and migrate apoplastically to endodermis from epidermis and accumulate in the vacuole; translocation of nanoparticles from roots to shoots was not observed (Li et al. 2016). In watermelon (*Citulluslanatus*) plants, TEM images helped to understand the uptake of magnetic iron oxide nanoparticles by root cells of plants, which further led to the analysis of the physiological changes brought about by the effects of nanoparticles (Wang et al. 2016). Yan et al. conducted an experiment to understand cesium retention and translocation under greenhouse conditions by applying stable cesium solution as droplets to the upper surface of soybean (Glycine max) trifoliate leaves (Yan et al. 2013). The presence of cesium on the leaf surface even after washing was confirmed with SEM coupled with EDX. Phytotoxicity studies with functionalized and non-functionalized SWCNTs on root elongation of different crop species was carried out by Canas et al. SEM images confirmed the presence of nanotube sheets on the surface of roots, but not in the inner root tissues (Canas et al. 2008). The effects of SWCNTs in the development of maize root tissue in association with changes in related gene expression was studied. TEM images of root tissues showed the presence of SWCNT bundles or aggregates in intercellular spaces (Yan et al. 2013). The toxic effects of MWCNTs on suspension rice cells were studied by culturing the nanotubes with cells. TEM images revealed the contact of each nanotube with the cell walls (Tan et al. 2009).

Scanning Transmission Electron Microscope (STEM) imaging provides the capabilities of both TEM and SEM. STEM imaging of the root nodules of ZnO nanoparticle-treated alfalfa (*Medicago sativa*) plants confirmed the bioaccumulation of ZnO nanoparticles in the root tissues. Plant tissues were harvested from different locations in order to understand the localization and accumulation of nanoparticles. In this study, imaging was carried out using BF/DF (Bright Field/Dark Field) Duo STEM, in which low magnification DF-STEM helped in locating the electron dense sample region due to presence of metal nanoparticles, whereas BF-STEM of the selected region revealed the presence of ZnO aggregates in cell walls and membrane. TEM images and STEM analysis of AgNP-treated *Arabidopsis thaliana* roots verified the presence of Ag^0 element in different compartments of root cells (Geisler-Lee et al. 2012). The uptake of TiO_2 NPs in hydroponically grown rice plants was also studied using STEM, in which the root and leaves of treated plants were sampled and analyzed. STEM analysis revealed accumulation of nanoparticles at root epidermal surface, and EDS analysis confirmed the presence of elemental Ti in the cytoplasm of treated roots (Deng et al. 2017). The major drawback of the electron microscopy analysis is that it provides information of only a small representative fraction of the plant area within a reasonable time period.

Figure 1 shows the presence of magnetic nanoparticles in pumpkin plants, which is confirmed with different imaging techniques such as Bright Field microscopy, Phase Contrast microscopy and TEM imaging (Corredor et al. 2009).

2.4 Autoradiography and PET/CT Imaging

The *in vivo* tracking of Copper-64 radiolabeled nanoparticles in lettuce (*Lactuca sativa*) was studied by Davis et al. (Davis et al. 2017). The aim of this study was to use radiolabeled NPs as a non-invasive analytical tool to quantitatively track and produce *in vivo* imaging of the translocation and accumulation of nanoparticles in lettuce plants by autoradiography and PET/CT and quantification by using a gamma counter. The biodistribution of radioactivity for each plant part is analyzed, in which most of the radioactivity was analyzed in roots. HPLC and gamma counting analysis confirmed that the radioactivity signals in plant parts was from [^{64}Cu]-NPs.

2.5 X-ray Computed Nanotomography (Nano-CT) and Dark-Field Microscopy Combined with Hyperspectral Imaging (DF-HIS)

Nano CT and DF-HIS are two complimentary imaging techniques that can be used to detect and visualize nanoparticles in plant tissue; by spectral confirmation with the 2D visualization DF-HIS tool and 3D imaging using nano-CT. DF-HIS can minimize the sample preparation and could detect and map the NP-specific reflectance spectra of the nanomaterial in complex environmental matrices in a

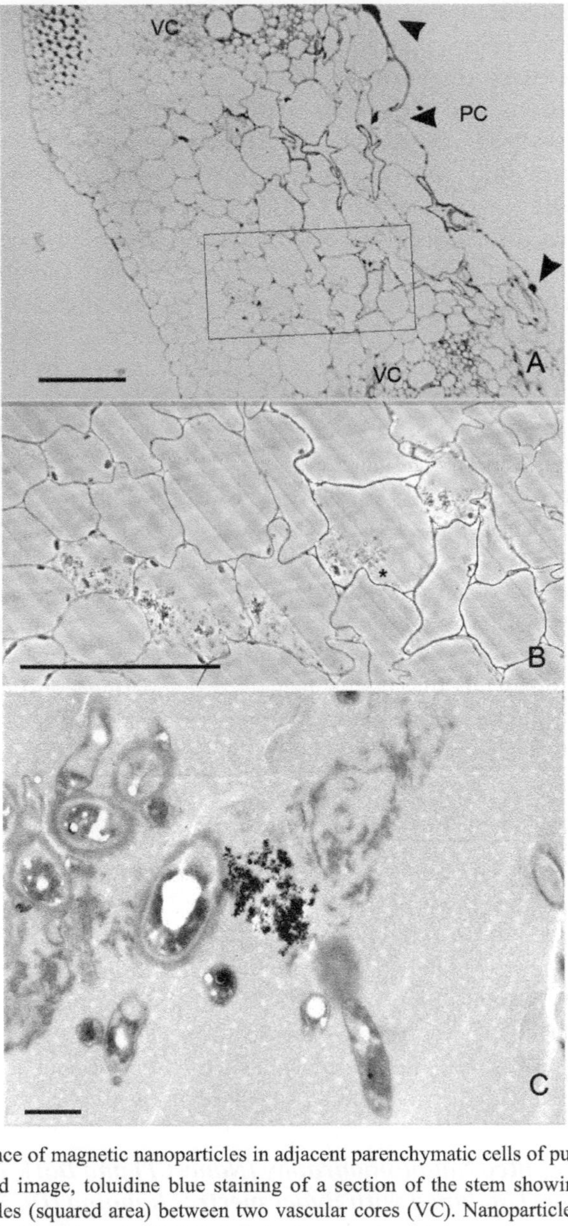

Fig. 1. Presence of magnetic nanoparticles in adjacent parenchymatic cells of pumpkin plants.

A. Bright field image, toluidine blue staining of a section of the stem showing cells carrying nanoparticles (squared area) between two vascular cores (VC). Nanoparticles also deposit in cell walls facing pith cavity (PC) (*labelled with arrow heads*).

B. Phase contrast image of a section consecutive to the one showed in (A) *with a detail of the region squared in (A)*. Dark material appears in the cell cytoplasm surrounded by dense structures. The surrounding cells do not show such a dense cytoplasm.

C. TEM image showing a detail of a nanoparticle aggregate inside one of the cells showed in (B) which is labelled by (*). Bar in A and B = 100 μm; C = 1 μm (*Adopted with permission from Reference 5*).

short time (minutes to hours) and narrow focus plane. Nano-CT imaging technique could provide valuable 3D information about plant-NP interaction. The technique is based on X-ray attenuation by the sample and it does not require any sectioning, labeling or staining of sample, thus, it can reduce the risk of artifacts from sample preparation. This combined method was successfully utilized by Avellan et al. to image negatively and positively charged gold nanoparticles in *Arabidopsis thaliana* upon their interaction with roots at the cellular level (Avellan et al. 2017). The technique provided direct evidence of accumulation of nanoparticles in detaching border-like cells (such as sheets of border cells separating from the root) and the associated mucilage could trap nanoparticles irrespective of their charge. This study also revealed charge specificity of border cells of root cap, in which they mainly adsorbed positively charged nanoparticles, thus preventing their further uptake and translocation; however, passage of negatively charged nanoparticles into the apoplast was allowed. This study presents direct evidence of the complimentary use of two emerging techniques, nano-CT and DF-HIS to assess the distribution of gold nanoparticles in plants.

3. Spectroscopy Methods

Inductively Coupled Plasma (ICP)-based spectroscopic techniques mainly include mass spectrometry (ICP-MS), Single Particle-Inductively Coupled Plasma-Mass Spectrometry (SP-ICP-MS) and Optical Emission Spectrometry (ICP-OES).

3.1 ICP-MS, SP-ICP-MS and ICP-OES

The principle of ICP includes the introduction of completely digested liquid sample into argon plasma as aerosols produced with the help of a nebulizer. The sample is ionized with the inductively coupled plasma, and the ions are separated, measured and quantified. Mass spectrometry (MS) and Optical Emission Spectrometry (OES) are the two widely used techniques for the detection and quantification of different elements. MS utilizes mass to charge ratio of particular element, whereas OES utilizes the electromagnetic radiation at wavelengths characteristic of a particular element. SP-ICP-MS (Single Particle-ICP-MS) is a developing technique for the simultaneous detection of size, size distribution and concentration of the studied particles. The technique can be used at very low nanoparticle concentrations in complex matrices.

Zhao et al. studied the genetic, morphological and physiological effects of Cu based nanopesticides on maize (*Zea mays*) plants (Zhao et al. 2017). Three week old maize plants were exposed to foliar application of different concentrations of $Cu(OH)_2$ nanopesticides. All the elements of dried leaves and root tissues of treated plants were analyzed using ICP-MS. Cu bioaccumulation on corn leaves was confirmed any analyzing the amount of Cu using ICP-MS. No change in Cu content was reported in plant roots which indicated that there was translocation of Cu from leaves to roots through phloem loading. The phytotoxicity of CeO_2 NPs on lettuce plants cultured in potted soil was studied and the Ce content in the roots and shoots of 30 day old treated plants was determined using ICP-MS (Gui et al. 2015).

SP-ICP-MS was used to analyze the uptake of gold nanoparticles by tomato (*Solanum lycopersicum* L.) plants. Specialized enzymatic digestion of plant tissues using Macerozyme R-10, a multicomponent enzyme mixture, was carried out to release nanoparticles from plant matrices followed by detection using SP-ICP-MS. Gold nanoparticles were detected even at very low concentration and hence this study reported an effective and rapid SP-ICP-MS analysis method for the direct and quantitative detection of nanoparticles in environmental samples (Dan et al. 2015) (Fig. 2). The internalization of AgNPs by *Arabidopsis* plants was also analyzed using SP-ICP-MS. Plants were exposed to 10 nm Ag NPs and size distribution of silver nanoparticles was determined using SP-ICP-MS by enzymatic digestion of root and shoot tissues (Bao et al. 2016). The detection of AgNP signals in shoot tissues confirmed the translocation of nanoparticles from root to shoots. Both these studies demonstrated that enzymatic digestion followed by SP-ICP-MS is a powerful tool for the quantitative determination of nanoparticles in plant tissues. SP-ICP-MS technology was also utilized to detect the size and size distribution of Ce nanoparticles, particle concentration and dissolved Ce in four different plant species: Cucumber (*Cucumis sativus*), tomato, soybean (*Glycine max*) and pumpkin (*Cucurbita pepo*) (Dan et al. 2016). The particle size detection limit was found to be 23–25 nm in this study. As in other studies with SP-ICP-MS, this study also showed the applicability of Macerozyme R-10 enzyme for the digestion of plant tissues and CeO_2 NP extraction without any dissolution of nanoparticles. Enzymatic digestion method was also utilized by Jimenez-Laman et al. for the extraction of Pt nanoparticles from different plant tissues (*Sinapis alba* and *Lapidium sativum*) without any change to their oxidation or aggregation state (Jiménez-Lamana et al. 2016). SP-ICP-MS results indicated the uptake and transportation of nanoparticles to the above ground portions and provided information about the size, size distribution and nanoparticle number concentration in different plant tissues. It also revealed the presence of nanoparticles in the dissolved and/or particulate form in the studied plants' tissues. The results suggested the use of plants to recover platinum from the environment. SP-ICP-MS was also used to access the size distribution and number

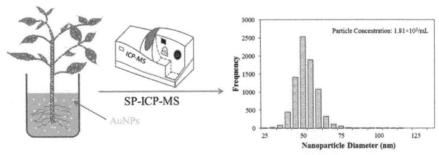

Fig. 2. SP-ICP-MS analysis of size, size distribution, particle concentration and dissolved Au concentration in tomato plant tissues (*Adopted with permission from Reference 43*).

concentration of Ti nanoparticles and bulk particles in rice plants. This study utilized two different extraction methods, microwave acid digestion and enzymatic digestion, in order to extract the particles from plant tissues. A difference in the size distribution of extracted TiO_2 particles was observed in rice plants exposed to TiO_2 NPs and TiO_2 bulk particles. The nanoparticle exposed plants showed a narrow size distribution whereas a broader distribution was found in bulk particle exposed plants. This study highlighted the ability of SP-ICP-MS to identify the difference in particle size distribution among nanoparticles and bulk particles (Deng et al. 2017). SP-ICP-MS was used to detect and quantify the presence of CuO NPs in edible plant tissues of lettuce (*Lactuca sativa* var. green leaf cultivar), kale (*Brassica oleracea,* var. *Acephala Lacinato*) and collard green (*Brassica oleracea,* var. *Acephala*) on exposure to nano-CuO (Keller et al. 2018). This tool serves in understanding the distribution of nanochemicals in agriculture that could help to design them more effectively with less risk to environment.

The concentration of Ag in sunflower (*Helianthus annus*) and sorghum (*Sorghum vulgare*) plants exposed to AgNPs was analyzed using ICP-OES (Pappas et al. 2017). In both plants, Ag was detected in all root samples with an increase in the concentration corresponding to an increase in the concentration of AgNPs in the soil. In *S. vulgare*, other than the roots, silver was not detected in any tissues except the leaf sample obtained from a particular concentration treatment group. In *H. annus*, roots and leaves of plants treated with all studied concentrations of AgNPs contained trace amounts of silver, and the stem samples of two highest concentration treated plants also contained silver traces. Deng et al. also utilized ICP-OES analytical method to analyze the total elemental concentration in rice plants after exposure to TiO_2 NPs and larger bulk particles (Deng et al. 2017). ICP-OES was successfully utilized in velvet mesquite (*Prosopis-juliflora-velutina*) plants to determine the concentration of Zn in different plant parts such as roots, stem and leaves upon treatment with ZnO nanoparticles (Hernandez-Viezcas et al. 2011). The seeds of alfalfa, corn, cucumber and tomato plants were treated with different concentrations of nano-ceria (CeO_2) and the presence of cerium in plant tissues was determined using ICP-OES (López-Moreno et al. 2010).

Laser Ablation-ICP-MS (LA-ICP-MS) is another powerful analytical technique that uses laser beam focused on the surface of sample to generate fine particles through a process called laser ablation. The ablated particles are then transported to the secondary excitation source of the ICP-MS instrument for digestion and ionization and the excited ions are further introduced to MS detector for elemental analysis. LA-ICP-MS was first used to image the distribution of gold nanoparticles (AuNPs) in earthworms (*Eiseniafetida*) (Unrine et al. 2010). The application of this technique in plant system for studying the uptake and distribution of AuNPs of different sizes (5, 10 and 15 nm) by tobacco (*Nicotiana tabacum* L. cv Xanthi) plants was reported by Judy et al. (Judy et al. 2010). LA-ICP-MS technique determined the presence of Au in the mesophyll of plant leaves. The team also studied the importance of particle size and surface coating on the bioavailability of AuNPs coated with tannate or citrate in tobacco (*Nicotiana tabacum* L. cv Xanthi) and wheat (*Triticum aestivum*). Plants were exposed to nanoparticles hydroponically for 3–7 days and the special distribution of Au in plant tissues was determined using LA-ICP-MS (Judy et al. 2012). Koelmel et al. studied the uptake and special distribution of differently charged gold nanoparticles (AuNPs) in rice roots and shoots under hydroponic conditions using LA-ICP-MS (Koelmel et al. 2013). The concentration of positively charged gold nanoparticles was found to be highest in plant roots and least in plant shoots, thus indicating preferential translocation. This study effectively used this technique to quantify AuNPs in rice plants. Hence, LA-ICP-MS can be effectively used to determine the uptake of insoluble nanoparticles by plant system.

3.2 GC-TOF MS (Gas Chromatography-Time of Flight-Mass Spectrometry)

Zhao et al. utilized GC-TOF-MS to analyze low molecular weight antioxidants produced as result of interaction of $Cu(OH)_2$ nanopesticides with corn plants (Zhao et al. 2017). Salunke et al. reported that GC-TOF-MS analysis was performed on ethyl acetate root extracts of the medicinal plant *Plumbago zeylanica* for the presence of flavonoids, sugar and organic acids, and nanoparticles such as silver (Ag), gold (Au) and bimetallic (Au-Ag) were successfully synthesized from the aqueous root extracts which were further utilized in biofilm control (Salunke et al. 2014).

4. Synchrotron Based Techniques

Synchrotron based imaging and spectroscopic methods have found great applications in plant science (Vijayan et al. 2015). Synchrotron radiation techniques are powerful tools used to study the chemical speciation, chemical transformation and distribution of metals in plants exposed to several metal and metal oxide nanoparticles. The advantages of synchrotron radiation includes little or no pretreatment of samples and the ability to identify the chemical form of the element of interest.

4.1 Synchrotron X-ray Imaging

Phase contrast images of water transport in the xylem vessels of living organisms were imaged using Synchrotron X-ray technique. In a study with gold nanoparticles, synchrotron X-ray imaging was used for the real time quantitative visualization of water conducting pathways and sap flows with the aim of investigating plant hydraulics (Hwang et al. 2014). The effects of AuNP solution on the water refilling function of xylem vessels was investigated in three monocots, maize, rice and bamboo (*Bambusoideae*), and compared with the water refilling speed with distilled water. It was observed that the average water refilling speed for both AuNP solution and distilled water were identical. However the advantage lies in the ability of AuNPs to be effectively used as flow tracers in xylem vessels during the initial time of 20–30 minutes without any physiological barrier. The particles gradually stained the surface of xylem cell walls so that their morphological structures could be easily distinguished in X-ray imaging. It was found that the interaction of AuNPs and stained xylem vessel structures was strong enough to protect the stained NPs from washing off with water.

4.2 Synchrotron X-ray Absorption Spectroscopy (XAS)

The XAS spectrum has two parts, X-ray absorption near edge structure (XANES) and Extended X-ray absorption fine structure (EXAFS). Gui et al. reported the speciation analysis of Ce in lettuce plants by XANES (Gui et al. 2015). The oxidation state of cerium in butter head lettuce root was identified with XANES and it was found that Ce in the roots existed in a mixed oxidation state of Ce (IV) and Ce (III). μXANES studies on root samples of CeO_2-treated corn plants revealed the speciation of Ce. The spectra obtained from epidermal and cortex spots were very similar to the spectra from CeO_2 NPs and data analysis reported that Ce was mainly present as Ce (IV) (Zhao et al. 2012). Duran et al. exploited XANES for the Cu chemical speciation analysis in *Phaseolus vulgaris* (common Brazilian bean) seeds upon exposure to Cu-based nanoparticle solution. Cu accumulation spots were located and chemical speciation of the element was determined (Duran et al. 2017). In velvet mesquite plants, XANES spectra was used to confirm that Zn was found in the Zn (II) form and not as nanoparticles upon treatment with ZnO nanoparticles (Hernandez-Viezcas et al. 2011). Lopez Moreno et al. studied the biotransformation of ZnO and CeO_2 nanoparticles in soybean seedlings and XAS spectra provided clear evidence of the presence of CeO_2 nanoparticles in roots, but not ZnO nanoparticles (López-Moreno et al. 2010). XANES spectra from model compounds provided information about the speciation of Zn and Ce. The speciation of Ce in four different plant species upon treating their roots with nano-ceria (4000 mg/L) was also determined based on XANES spectra for Ce from roots of treated plants on comparison with CeO_2 nanoparticles as model compounds. Ce was found to be in the same oxidation state in roots and, thus, confirmed that no chemical transformation has happened in plant tissues after uptake of CeO_2 nanoparticles (López-Moreno et al. 2010).

4.3 μ-XRF (micro-X-ray Fluorescence Spectroscopy)

μ-XRF is a non-destructive microanalytical method used to evaluate the special distribution of elements in biological samples. The distribution of Ce in the root tissues of corn plants was mapped using μ-XRF. Ce intensity was found to be higher in the root epidermis, whereas low signals were detected in the central cylinder of the root tissues which confirmed the presence of more Ce in the vascular tissues, justifying the confocal fluorescence results presented earlier (Zhao et al. 2012). Larue et al. used μ-XRF based on synchrotron radiation in order to evaluate the distribution of Ti in roots and leaves of wheat and rapeseed plants. TiO_2 NP accumulation was reported in both plantlets upon root and leaf exposure, and the Ti content was found to be greater in rapeseed than in wheat (Larue et al. 2012). In *Phaseolus vulgaris* seeds, the mapping of Cu accumulation spots was executed with μ-XRF and this helps in selecting proper seed region for XANES analysis (Duran et al. 2017). The μ-XRF maps identified three different regions for speciation studies: Outside of seed coat, inside of seed coat and cotyledon near the seed coat. In the studies of growth enhancement of bean seedlings using magnetite nanoparticles, μ-XRF and X-ray tomography results showed that Fe provided as nanoparticle solution could penetrate the seeds (Duran et al. 2018). μ-XRF studies were carried out to understand the role of chelants on Cu distribution and speciation in rye grass (*Loliummultiflorum*). μ-XRF mapping revealed that the chelating agent, EEDA (Ethylenediamine disuccinic acid), alleviated the deposition of Cu at root meristem of root apex and lateral root junction, however, it facilitated the Cu transport to root stele and further upward translocation (Zhao et al. 2018). Upon treatment with ZnO nanoparticles in velvet mesquite plants, μ-XRF analysis confirmed the presence of Zn in vascular system of roots and leaves (Hernandez-Viezcas et al. 2011) (Fig. 3).

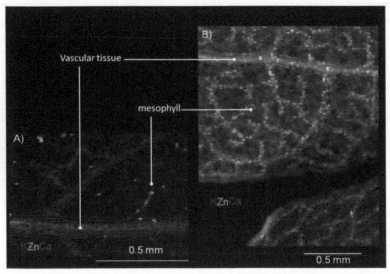

Fig. 3. XRF maps of velvet mesquite plant leaves (A) Control leaves (B) Leaves treated with 1000 mg/L ZnO nanoparticles (*Red color-Potassium, Green color-Zinc, and Purple color-Calcium*) (*Adopted with permission from Reference 49*).

Other Techniques

Khodakovskaya et al. introduced an integrated genomic and photothermal based analysis in order to understand the interaction of carbon-based nanomaterials with tomato plants. A photothermal (PT) and photoacoustic mapping of multi walled carbon nanotubes (MWCNTs) in tomato roots, leaves and fruits were done. This technique is based on non-irradiative conversion of absorbed laser energy into thermal and acoustic phenomena with high sensitivity and resolution in deeper tissues under *in vivo* condition. PT/PA scanning cytometry platform was developed on the basis of an inverted microscope and the researchers described the whole set up in detail in the publication (Khodakovskaya et al. 2011). PT/PA scanning cytometry demonstrated the presence of CNTs in both roots and leaves of treated tomato plants. Microarray technology was utilized to study the altered gene expression in tissues that occurred due to the external stress associated with the nanotube update by plants, as demonstrated by PT/PA scanning cytometry. Raman spectroscopy is one of the most sensitive, non-destructive methods that analyze the presence of nanomaterials in plant cells and organs. In another study, Khodakovskaya et al. reported that Raman scattering detected the presence of CNT aggregates in flowers of tomato plants grown in soil treated with CNTs (Khodakovskaya et al. 2013). The confirmation of the presence of nanomaterials in reproductive parts raised the safety concerns about the use of nanomaterials as growth regulators in plants. Micro Particle Induced X-ray Emission (μ-PIXE) is another non-destructive elemental localization and quantification analysis technique. Larue et al. used μ-PIXE along with other complementary techniques, such as SEM, TEM and μ-XRF, in order to assess the accumulation of nanoparticles in wheat (*Triticum aestivum*), oilseed rape (*Brassica napus*) and *Arabidopsis thaliana* upon being exposed to TiO_2 nanoparticles hydroponically (Larue et al. 2011). The physiological effects of TiO_2 nanoparticles on plants were analyzed in this study. The technique was also utilized by the same group in other nanotoxicity study in wheat plants in order to confirm the uptake and accumulation of nanoparticles in plant parts (Larue et al. 2012).

Conclusion

The physical and chemical properties of nanomaterials often differ from those of bulk materials and they require specialized risk assessment detection and visualization of nanoparticles in environmental samples, including crop plants. The major challenge in the detection of nanoparticles lies in their extremely small size and the lack of proper qualitative and quantitative methods which are still being developed. Novel tools are required in order to study the interaction of nanoparticles and analyze specific plant responses to such interaction. Most of the techniques are offline and require complicated sample preparation methods. Recent developments in imaging and spectroscopy methods provide complementary insights into the bioaccumulation behavior of nanoparticles within plant system. Electron microscopy methods have improved with the development of STEM and High Resolution TEM (HR-TEM), which again can be combined with powerful analytical techniques, such as various ICP methods and X-ray analysis techniques.

Such complementary methods will provide more powerful information and understanding of the interaction of nanomaterials with plant systems, which could lead to great advancements in agriculture.

References

Abd-Alla, M.H., N.A. Nafady and D.M. Khalaf. 2016. Assessment of silver nanoparticles contamination on faba bean-*Rhizobium leguminosarum* bv. viciae-Glomus aggregatum symbiosis: Implications for induction of autophagy process in root nodule. Agriculture, Ecosystems & Environment 218: 163–77.

Abraham, P.M., S. Barnikol, T. Baumann, M. Kuehn, N.P. Ivleva and G.E. Schaumann. 2013. Sorption of silver nanoparticles to environmental and model surfaces. Environmental Science & Technology 47(10): 5083–91.

Avellan, A., F. Schwab, A. Masion, P. Chaurand, D. Borschneck, V. Vidal, J. Rose, C. Santaella and C. Levard. 2017. Nanoparticle uptake in plants: Gold nanomaterial localized in roots of *arabidopsis thaliana* by X-ray computed nanotomography and hyperspectral imaging. Environmental Science & Technology 51(15): 8682–91.

Bao, D., Z.G. Oh and Z. Chen. 2016. Characterization of silver nanoparticles internalized by Arabidopsis plants using single particle ICP-MS analysis. Frontiers in Plant Science 7: 32.

Begum, P. and B. Fugetsu. 2012. Phytotoxicity of multi-walled carbon nanotubes on red spinach (*Amaranthus tricolor* L.) and the role of ascorbic acid as an antioxidant. Journal of Hazardous Materials 243: 212–22.

Canas, J.E., M. Long, S. Nations, R. Vadan, L. Dai, M. Luo, R. Ambikapathi, E.H. Lee and D. Olszyk. 2008. Effects of functionalized and nonfunctionalized single-walled carbon nanotubes on root elongation of select crop species. Environmental Toxicology and Chemistry 27(9): 1922–31.

Chang, K.C., Y.W. Chiang, C.H. Yang and J.W. Liou. 2012. Atomic force microscopy in biology and biomedicine. Tzu Chi Medical Journal 24(4): 162–9.

Corredor, E., P.S. Testillano, M.J. Coronado, P. González-Melendi, R. Fernández-Pacheco, C. Marquina, M.R. Ibarra, J.M. de la Fuente, D. Rubiales, A. Pérez-de-Luque and M.C. Risueño. 2009. Nanoparticle penetration and transport in living pumpkin plants: *In situ* subcellular identification. BMC Plant Biology (1): 45.

Dan, Y., W. Zhang, R. Xue, X. Ma, C. Stephan and H. Shi. 2015. Characterization of gold nanoparticle uptake by tomato plants using enzymatic extraction followed by single-particle inductively coupled plasma–mass spectrometry analysis. Environmental Science & Technology 49(5): 3007–14.

Dan, Y., X. Ma, W. Zhang, K. Liu, C. Stephan and H. Shi. 2016. Single particle ICP-MS method development for the determination of plant uptake and accumulation of CeO_2 nanoparticles. Analytical and Bioanalytical Chemistry 408(19): 5157–67.

Davis, R.A., D.A. Rippner, S.H. Hausner, S.J. Parikh, A.J. McElrone and J.L. Sutcliffe. 2017. *In vivo* tracking of copper-64 radiolabeled nanoparticles in *Lactuca sativa*. Environmental Science & Technology 51(21): 12537–46.

Deng, Y., E.J. Petersen, K.E. Challis, S.A. Rabb, R.D. Holbrook, J.F. Ranville, B.C. Nelson and B. Xing. 2017. Multiple method analysis of TiO_2 nanoparticle uptake in rice (*Oryza sativa* L.) plants. Environmental Science & Technology 51(18): 10615–23.

Duran, N.M., S.M. Savassa, R.G. Lima, E. de Almeida, F.S. Linhares, C.A. van Gestel and H.W. Pereira de Carvalho. 2017. X-ray spectroscopy uncovering the effects of Cu-based nanoparticle concentration and structure on *Phaseolus vulgaris* germination and seedling development. Journal of Agricultural and Food Chemistry 65(36): 7874–84.

Duran, N.M., M. Medina-Llamas, J.G. Cassanji, R.G. de Lima, E. de Almeida, W.R. Macedo, D. Mattia and H.W. Pereira de Carvalho. 2018. Bean seedling growth enhancement using magnetite nanoparticles. Journal of Agricultural and Food Chemistry. doi:10.1021/acs.jafc.8b00557 2018 May 25.

Fan, R., Y.C. Huang, M.A. Grusak, C.P. Huang and D.J. Sherrier. 2014. Effects of nano-TiO_2 on the agronomically-relevant Rhizobium–legume symbiosis. Science of the Total Environment 466: 503–12.

Geisler-Lee, J., Q. Wang, Y. Yao, W. Zhang, M. Geisler, K. Li, Y. Huang, Y. Chen, A. Kolmakov and X. Ma. 2012. Phytotoxicity, accumulation and transport of silver nanoparticles by *Arabidopsis thaliana*. Nanotoxicology 7(3): 323–37.

Gogos, A., J. Moll, F. Klingenfuss, M. Heijden, F. Irin, M.J. Green, R. Zenobi and T.D. Bucheli. 2016. Vertical transport and plant uptake of nanoparticles in a soil mesocosm experiment. Journal of Nanobiotechnology 14(1): 40.

González-Melendi, P., R. Fernández-Pacheco, M.J. Coronado, E. Corredor, P.S. Testillano, M.C. Risueno, C. Marquina, M.R. Ibarra, D. Rubiales and A. Pérez-de-Luque. 2007. Nanoparticles as smart treatment-delivery systems in plants: Assessment of different techniques of microscopy for their visualization in plant tissues. Annals of Botany 101(1): 187–95.

Gui, X., Z. Zhang, S. Liu, Y. Ma, P. Zhang, X. He, Y. Li, J. Zhang, H. Li, Y. Rui and L. Liu. 2015. Fate and phytotoxicity of CeO_2 nanoparticles on lettuce cultured in the potting soil environment. PloS One 10(8): e0134261.

Hernandez-Viezcas, J.A., H. Castillo-Michel, A.D. Servin, J.R. Peralta-Videa and J.L. Gardea-Torresdey. 2011. Spectroscopic verification of zinc absorption and distribution in the desert plant *Prosopis juliflora-velutina* (velvet mesquite) treated with ZnO nanoparticles. Chemical Engineering Journal 170(2-3): 346–52.

Hischemöller, A., J. Nordmann, P. Ptacek, K. Mummenhoff and M. Haase. 2009. *In vivo* imaging of the uptake of up conversion nanoparticles by plant roots. Journal of Biomedical Nanotechnology 5(3): 278–84.

Huang, Y.C., R. Fan, M.A. Grusak, J.D. Sherrier and C.P. Huang. 2014. Effects of nano-ZnO on the agronomically relevant Rhizobium–legume symbiosis. Science of the Total Environment 497: 78–90.

Hwang, B.G., S. Ahn and S.J. Lee. 2014. Use of gold nanoparticles to detect water uptake in vascular plants. PloS One 9(12): e114902.

Jiménez-Lamana, J., J. Wojcieszek, M. Jakubiak, M. Asztemborska and J. Szpunar. 2016. Single particle ICP-MS characterization of platinum nanoparticles uptake and bioaccumulation by *Lepidium sativum* and *Sinapis alba* plants. Journal of Analytical Atomic Spectrometry 31(11): 2321–9.

Judy, J.D., J.M. Unrine and P.M. Bertsch. 2010. Evidence for biomagnification of gold nanoparticles within a terrestrial food chain. Environmental Science & Technology 45(2): 776–81.

Judy, J.D., J.M. Unrine, W. Rao, S. Wirick and P.M. Bertsch. 2012. Bioavailability of gold nanomaterials to plants: Importance of particle size and surface coating. Environmental Science & Technology 46(15): 8467–74.

Keller, A.A., Y. Huang and J. Nelson. 2018. Detection of nanoparticles in edible plant tissues exposed to nano-copper using single-particle ICP-MS. Journal of Nanoparticle Research 20(4): 101.

Khodakovskaya, M.V., B.S. Kim, J.N. Kim, M. Alimohammadi, E. Dervishi, T. Mustafa and C.E. Cernigla. 2013. Carbon nanotubes as plant growth regulators: Effects on tomato growth, reproductive system and soil microbial community. Small 9(1): 115–23.

Khodakovskaya, M.V., K. de Silva, D.A. Nedosekin, E. Dervishi, A.S. Biris, E.V. Shashkov, E.I. Galanzha and V.P. Zharov. 2011. Complex genetic, photothermal and photoacoustic analysis of nanoparticle-plant interactions. Proceedings of the National Academy of Sciences 108(3): 1028–33.

Koelmel, J., T. Leland, H. Wang, D. Amarasiriwardena and B. Xing. 2013. Investigation of gold nanoparticles uptake and their tissue level distribution in rice plants by laser ablation-inductively coupled-mass spectrometry. Environmental Pollution 174: 222–8.

Lahiani, M.H., J. Chen, F. Irin, A.A. Puretzky, M.J. Green and M.V. Khodakovskaya. 2015. Interaction of carbon nanohorns with plants: Uptake and biological effects. Carbon 81: 607–19.

Larue, C., G. Veronesi, A.M. Flank, S. Surble, N. Herlin-Boime and M. Carrière. 2012. Comparative uptake and impact of TiO_2 nanoparticles in wheat and rapeseed. Journal of Toxicology and Environmental Health, Part A 75(13-15): 722–34.

Larue, C., H. Khodja, N. Herlin-Boime, F. Brisset, A.M. Flank, B. Fayard, S. Chaillou and M. Carrière. 2011. Investigation of titanium dioxide nanoparticles toxicity and uptake by plants. In Journal of Physics: Conference Series 2011 (Vol. 304, No. 1, p. 012057). IOP Publishing.

Larue, C., J. Laurette, N. Herlin-Boime, H. Khodja, B. Fayard, A.M. Flank, F. Brisset and M. Carriere. 2012. Accumulation, translocation and impact of TiO$_2$ nanoparticles in wheat (*Triticum aestivum* spp.): Influence of diameter and crystal phase. Science of the Total Environment 431: 197–208.

Larue, C., M. Pinault, B. Czarny, D. Georgin, D. Jaillard, N. Bendiab, M. Mayne-L'Hermite, F. Taran, V. Dive and M. Carrière. 2012. Quantitative evaluation of multi-walled carbon nanotube uptake in wheat and rapeseed. Journal of Hazardous Materials 227: 155–63.

Lenaghan, S.C. and M. Zhang. 2012. Real-time observation of the secretion of a nanocomposite adhesive from English ivy (*Hedera helix*). Plant Science 183: 206–11.

Li, J., J. Hu, C. Ma, Y. Wang, C. Wu, J. Huang and B. Xing. 2016. Uptake, translocation and physiological effects of magnetic iron oxide (γ-Fe$_2$O$_3$) nanoparticles in corn (*Zea mays* L.). Chemosphere 159: 326–34.

Lin, D. and B. Xing. 2008. Root uptake and phytotoxicity of ZnO nanoparticles. Environmental Science & Technology 42(15): 5580–5.

Lin, S., J. Reppert, Q. Hu, J.S. Hudson, M.L. Reid, T.A. Ratnikova, A.M. Rao, H. Luo and P.C. Ke. 2009. Uptake, translocation and transmission of carbon nanomaterials in rice plants. Small 5(10): 1128–32.

Liu, Q., B. Chen, Q. Wang, X. Shi, Z. Xiao, J. Lin and X. Fang. 2009. Carbon nanotubes as molecular transporters for walled plant cells. Nano Letters 9(3): 1007–10.

López-Moreno, M.L., G. de la Rosa, J.Á. Hernández-Viezcas, H. Castillo-Michel, C.E. Botez, J.R. Peralta-Videa and J.L. Gardea-Torresdey. 2010. Evidence of the differential biotransformation and genotoxicity of ZnO and CeO$_2$ nanoparticles on soybean (Glycine max) plants. Environmental Science & Technology 44(19): 7315–20.

López-Moreno, M.L., G. de la Rosa, J.A. Hernández-Viezcas, J.R. Peralta-Videa and J.L. Gardea-Torresdey. 2010. XAS corroboration of the uptake and storage of CeO$_2$ nanoparticles and assessment of their differential toxicity in four edible plant species. Journal of Agricultural and Food Chemistry 58(6): 3689.

Ma, X., J. Geiser-Lee, Y. Deng and A. Kolmakov. 2010. Interactions between engineered nanoparticles (ENPs) and plants: Phytotoxicity, uptake and accumulation. Science of the Total Environment. 408(16): 3053–61.

Marchiol, L., A. Mattiello, F. Pošćić, G. Fellet, C. Zavalloni, E. Carlino and R. Musetti. 2016. Changes in physiological and agronomical parameters of barley (*Hordeum vulgare*) exposed to cerium and titanium dioxide nanoparticles. International Journal of Environmental Research and Public Health 13(3): 332.

Milewska-Hendel, A., M. Zubko, J. Karcz, D. Stróż and E. Kurczyńska. 2017. Fate of neutral-charged gold nanoparticles in the roots of the *Hordeum vulgare* L. cultivar Karat. Scientific Reports. 7(1): 3014.

Nair, R., A.C. Poulose, Y. Nagaoka, Y. Yoshida, T. Maekawa and D.S. Kumar. 2011. Uptake of FITC labeled silica nanoparticles and quantum dots by rice seedlings: Effects on seed germination and their potential as biolabels for plants. Journal of Fluorescence 21(6): 2057.

Navarro, D.A., M.A. Bisson and D.S. Aga. 2012. Investigating uptake of water-dispersible CdSe/ZnS quantum dot nanoparticles by *Arabidopsis thaliana* plants. Journal of Hazardous Materials 211: 427–35.

Ndeh, N.T., S. Maensiri and D. Maensiri. 2017. The effect of green synthesized gold nanoparticles on rice germination and roots. Advances in Natural Sciences: Nanoscience and Nanotechnology 8(3): 035008.

Pappas, S., U. Turaga, N. Kumar, S. Ramkumar and R.J. Kendall. 2017. Effect of Concentration of Silver Nanoparticles on the uptake of Silver from Silver Nanoparticles in soil. International Journal of Environmental and Agriculture Research 3: 80–90.

Pariona, N., A.I. Martinez, H.M. Hdz-García, L.A. Cruz and A. Hernandez-Valdes. 2017. Effects of hematite and ferrihydrite nanoparticles on germination and growth of maize seedlings. Saudi Journal of Biological Sciences 24(7): 1547–54.

Salunke, G.R., S. Ghosh, R.S. Kumar, S. Khade, P. Vashisth, T. Kale, S. Chopade, V. Pruthi, G. Kundu, J.R. Bellare and B.A. Chopade. 2014. Rapid efficient synthesis and characterization of silver, gold and bimetallic nanoparticles from the medicinal plant *Plumbago zeylanica* and their application in biofilm control. International Journal of Nanomedicine 9: 2635.

Siani, N.G., S. Fallah, L.R. Pokhrel and A. Rostamnejadi. 2017. Natural amelioration of zinc oxide nanoparticle toxicity in fenugreek (*Trigonellafoenum-gracum*) by arbuscular mycorrhizal (*Glomus intraradices*) secretion of glomalin. Plant Physiology and Biochemistry 112: 227–38.

Tan, X.M., C. Lin and B. Fugetsu. 2009. Studies on toxicity of multi-walled carbon nanotubes on suspension rice cells. Carbon 47(15): 3479–87.

Unrine, J.M., S.E. Hunyadi, O.V. Tsyusko, W. Rao, W.A. Shoults-Wilson and P.M. Bertsch. 2010. Evidence for bioavailability of Au nanoparticles from soil and biodistribution within earthworms (*Eiseniafetida*). Environmental Science & Technology 44(21): 8308–13.

Vijayan, P., I.R. Willick, R. Lahlali, C. Karunakaran and K.K. Tanino. 2015. Synchrotron radiation sheds fresh light on plant research: The use of powerful techniques to probe structure and composition of plants. Plant and Cell Physiology 56(7): 1252–63.

Wang, P., N.W. Menzies, E. Lombi, B.A. McKenna, S. James, C. Tang and P.M. Kopittke. 2015. Synchrotron-based X-ray absorption near-edge spectroscopy imaging for laterally resolved speciation of selenium in fresh roots and leaves of wheat and rice. Journal of Experimental Botany 66(15): 4795–806.

Wang, P., E. Lombi, F.J. Zhao and P.M. Kopittke. 2016. Nanotechnology: A new opportunity in plant sciences. Trends in Plant Science 21(8): 699–712.

Wang, Y., J. Hu, Z. Dai, J. Li and J. Huang. 2016. *In vitro* assessment of physiological changes of watermelon (*Citrulluslanatus*) upon iron oxide nanoparticles exposure. Plant Physiology and Biochemistry 108: 353–60.

Wild, E. and K.C. Jones. 2009. Novel method for the direct visualization of *in vivo* nanomaterials and chemical interactions in plants. Environmental Science & Technology 43(14): 5290–4.

Wu, H., I. Santana, J. Dansie and J.P. Giraldo. 2017. *In vivo* delivery of nanoparticles into plant leaves. Current Protocols in Chemical Biology 14: 269–84.

Yan, D., Y. Zhao, A. Lu, S. Wang, D. Xu and P. Zhang. 2013. Effects of accompanying anions on cesium retention and translocation via droplets on soybean leaves. Journal of Environmental Radioactivity 126: 232–8.

Yan, S., L. Zhao, H. Li, Q. Zhang, J. Tan, M. Huang, S. He and L. Li. 2013. Single-walled carbon nanotubes selectively influence maize root tissue development accompanied by the change in the related gene expression. Journal of Hazardous Materials 246: 110–8.

Yang, X., H. Pan, P. Wang and F.J. Zhao. 2017. Particle-specific toxicity and bioavailability of cerium oxide (CeO$_2$) nanoparticles to *Arabidopsis thaliana*. Journal of Hazardous Materials 322: 292–300.

Zhao, L., J.P. Peralta-Videa, A. Varela-Ramirez, H. Castillo-Michel, C. Li, J. Zhang, R.J. Aguilera, A.A. Keller and J.L. Gardea-Torresdey. 2012. Effect of surface coating and organic matter on the uptake of CeO$_2$ NPs by corn plants grown in soil: Insight into the uptake mechanism. Journal of Hazardous Materials 225: 131–8.

Zhao, L., Q. Hu, Y. Huang and A.A. Keller. 2017. Response at genetic, metabolic, and physiological levels of maize (Zea mays) exposed to a Cu(OH)$_2$ nanopesticide. ACS Sustainable Chemistry & Engineering 5(9): 8294–301.

Zhao, Y.P., J.L. Cui, T.S. Chan, J.C. Dong, D.L. Chen and X.D. Li. 2018. Role of chelant on Cu distribution and speciation in *Loliummultiflorum* by synchrotron techniques. Science of the Total Environment 621: 772–81.

CHAPTER 3

Bacterial Synthesis of Metallic Nanoparticles

Current Trends and Potential Applications as Soil Fertilizer and Plant Protectants

Shweta Agrawal,[1,]* *Mrinal Kuchlan,*[2] *Jitendra Panwar*[3]
and *Mahaveer Sharma*[2]

1. Introduction

The rapid development of nanotechnology has resulted in the production and utilization of a number of nanoparticles of varying composition, size and shape characteristics (Vankova et al. 2017). NPs are metal and metal oxide particles having at least one dimension in the size range of 1–100 nm exhibiting different shapes, such as spherical, triangular, rod, etc. Nano-size provides unique chemical, physical and optical properties to these NPs as compared to the corresponding bulk material and, therefore, synthesis of NPs is one of the challenging areas of current research (Gade et al. 2010, Ramanathan et al. 2010, Rai and Ingle 2012, Bradfield et al. 2017, Javed et al. 2017). Progress in the field of nanotechnology has been rapid, especially in the development of novel NP synthesis protocols and their characterization techniques (Gade et al. 2010, Ramanathan et al. 2010,

[1] Shri Vaishnav Institute of Science, Shri Vaishnav Vidyapeeth Vishwavidyalaya, Indore 453111, Madhya Pradesh, India.
[2] ICAR-Indian Institute of Soybean Research, Khandwa Road, Indore 452 001, Madhya Pradesh, India.
[3] Department of Biological Sciences, Birla Institute of Technology & Science (BITS), Pilani-333 031, Rajasthan, India.
* Corresponding author: shweta.agrawal24@gmail.com

Rai and Ingle 2012). Recently, there has been tremendous excitement in the study and synthesis of nanoparticles using some natural biological systems (Bhargava et al. 2013). This has led to the development of various biomimetic approaches for the growth of advanced nanomaterials (Jain et al. 2013, Bhargava et al. 2016). Microorganisms, such as bacteria, yeast and fungi, are known to produce inorganic materials either intra- or extracellularly (Nangia et al. 2009).

These biomimetic approaches are greener as compared to the synthetic methods, since no toxic chemicals are involved in the synthesis of NPs under low energy consumption, i.e., using ambient conditions of temperature, pH and pressure. Moreover, the microbial cultures are easy to handle, simple in downstream processing of microbial biomass and have ease of scalability and generate highly stable NPs, making the microbial synthesis approach more advantageous. Additionally, the particles generated by microbial methods have higher catalytic reactivity and greater specific surface area, as well as an improved affinity and compatibility between the enzyme and metal salts owing to the bacterial carrier matrix. Hence, an enormous amount of research efforts are currently being diverted towards the microbial synthesis of NPs (Pugazhenthiran et al. 2009, Ramanathan et al. 2010, Li et al. 2011, Rai et al. 2011, Jayaseelan 2012). Bacteria are considered as a potential biofactory for the synthesis of NPs like gold, silver, platinum, palladium, titanium, titanium dioxide, magnetite, cadmium sulphide and so forth (Iravani 2014).

Demand for sustaining higher agricultural production will continue to be constrained by a number of factors, viz., continuous mining of mineral nutrients (nutrient deficiency), soil pollution due to the use of high doses of chemical pesticides and fertilizers, recurrent droughts and biotic factors (insect pests, diseases, weeds). Nanotechnology has the potential to sustain the productivity of plants and soil of the agricultural sector and serve as an advanced tool for the delivery of the nanocides; slow release of biofertilizers, nanomaterial-assisted fertilizers as well as micronutrients for efficient use; and field applications of agrochemicals, nanomaterial-assisted delivery of genetic material for crop improvement. In the near future, nano-based catalysts which can be used to increase the efficiency of pesticides and herbicides at much lower doses will also be available. Thus, the application of nanotechnology in agriculture could prove to be a great boon through exploration and revisiting of upcoming products and techniques. Very recently, it has been reported that nearly 14,000 documents and 2707 patents related to applications of nanotechnology in food or agriculture have been published, indicating high prospects of applied research in this field (Rai and Ingle 2012, Ghormade et al. 2012, Prasad et al. 2017).

This chapter provides a brief scenario of the current research activities concentrated on the synthesis of NPs by different bacteria, NP biosynthesis mechanisms and current applications of biosynthesized NPs in various agricultural sectors. Moreover, the chapter also provides possible considerations and a future road map for developing the potential of NP synthesis by bacteria.

2. Synthesis of Nanoparticles by Bacteria

There have been many reports where a variety of microorganisms have been used for producing metallic NPs of different shapes and natures under different conditions. We hereby provide an overview on bacteria mediated NP synthesis and discuss a few landmark reports. A detailed list of reports on synthesis of various NPs by different bacteria, the synthesis parameters and characteristics of biosynthesized nanoparticles is provided in Table 1.

2.1 Silver Nanoparticles (AgNPs)

In the era of nanotechnology, the nanoform of silver, i.e., silver nanoparticles (AgNPs), is gaining noteworthy considerations due to its unique properties. Compared to bulk silver, AgNPs, due to their miniscule size, possess a large surface-area-to-volume ratio and, thus, exhibit a high reaction rate that helps in establishing a better contact with microorganisms, thereby enhancing their antimicrobial property, making them a potent broad-spectrum antimicrobial agent. The effectiveness of AgNPs could be estimated by the fact that they kill approximately 650 types of pathogenic microbes, such as bacteria, fungi, viruses, yeasts, etc., and kill pathogenic bacteria in a very short time, i.e., 30 min (Kathiresan et al. 2010, Mishra and Singh 2014). The synthesis of AgNPs by *Pseudomonas stutzeri* AG259 was reportedly the first report on the synthesis of NPs by microorganisms. The AgNPs, formed as single crystals, were embedded in the organic matrix of the bacteria, growing up to 200 nm in size with equilateral triangle and hexagonal forms (Duran et al. 2011). In an exciting study on *Lactobacillus* strains present in buttermilk, a bacterium which does not normally encounter large metal ion concentrations, when exposed to silver and gold ions, led to large-scale production and accumulation of metal NPs within the bacterial cells (Sastry et al. 2003). Contrastingly, marine microbes, which have existed on the sea bed, have capabilities to reduce inorganic elements and thereby are reported to synthesize NPs. Intracellular AgNPs were reported to be synthesized by a marine, silver-tolerant bacterium, *Idiomarina* sp. PR58-8, *Oscillatoria willei* and *Stenottrophomonas* sp. (Singh et al. 2015). Novel approaches using a combination of culture supernatanant of *Bacillus subtilis* and microwave irradiation in water, in the absence of a surfactant or soft template, lead to the formation of AgNPs of dimension 5–60 nm (Saifuddin et al. 2009). Similarly, solar intensity of 70,000 lx, $AgNO_3$ concentration of 3 mg/mL and 2 mM NaCl content led to the conversion of 98.23 ± 0.06% of the Ag^+ (1 mM) AgNPs within 80 min (Wei et al. 2012).

The interface between endophytes and nanomaterials is a relatively new and unexplored area. An unidentified endophytic bacterium isolated from the leaf and stem segments of *Coffee arabica* L. (Baker and Shreedharmurthy 2012) and endophytic *Bacillus cereus* isolated from the *Garcinia xanthochymus* (Sunkar and Nachiyar 2012) have been found capable of producing extracellular synthesis of

Table 1. Synthesis of metal nanoparticle by bacteria.

Name of Strain (Alphabetically Arrange on the Basis of Name of Bacteria)	Conditions (Culture, Metal Salt, Temperature (°C), pH, Incubation Time (h), Agitation (rpm) and Illumination)	Shape, Size (nm) and Location of NPs	References
Silver Nanoparticles			
Escherichia coli AUCAS 112	CFS[1], silver nitrate, 25, NA[2], 72, NA, dark	Spherical, 5–20, Ex[3]	(Kathiresan et al. 2010)
Klebsiella pneumoniae	CFS, silver nitrate, NA, NA, 5 min, NA, light	NA, 5 to 32 nm, Ex	(Shahverdi et al. 2007)
Bacillus flexus S-27	CFS, silver nitrate, RT, NA, NA, NA, NA	Spherical, 12–14 and triangular 61, Ex	(Priyadarshini et al. 2013)
Bacillus sp.	Cell culture, silver nitrate, NA, NA, 7 d, NA, dark	NA, 10-15 nm, In[4]	(Pugazhenthiran et al. 2009)
Lactobacillus casei subsp. casei	Cell mass, silver nitrate, 25 (6.5, 7 and 7.5), NA, 80, NA, glucose as electron donor	spherical, single (25–50 nm) or in aggregates (100 nm), inside and outside of the cells	(Korbekandi et al. 2012)
Pseudomonas stutzeri AG259	Cell culture, silver nitrate, 30, 48, dark	equilateral triangles and hexagons, 200, cell poles	(Klaus et al. 1999)
Pseudomonas aeruginosa (JQ989348)	CFS, silver nitrate, 30, NA, 8, shaker, NA	Spherical, 13 to 76, Ex	(Ramalingam et al. 2014)
Plectonema boryanum UTEX	cell culture, silver nitrate, 25–100, NA, 28 d, NA, NA		(Lengke et al. 2007)
Streptomyces sp. BDUKAS10	CFS, silver nitrate, 32, NA, 36, dark	Spherical, 21–48 nm, filtrate	(Sivalingam et al. 2012)
Pseudomonas aeruginosa	CFS, silver nitrate, NA, NA, 72, NA, illuminated	NA, 20–100 nm, filtrate	(Jeevan et al. 2012)
Calothrix pulvinata ALCP 745A	Cell culture, AgNO$_3$, NA, NA, 24, NA, NA	NA, 15 nm inside heterocyst and 10 nm extracellular NPs	(Brayner et al. 2007)
Anabaena flos-aquae ALCP B24	Cell culture, AgNO$_3$, NA, NA, 0.5, NA, NA	Spherical, 40 nm, In and 25 nm, Ex	(Brayner et al. 2007)

Table 1 contd. ...

...Table 1 contd.

Name of Strain (Alphabetically Arrange on the Basis of Name of Bacteria)	Conditions (Culture, Metal Salt, Temperature (°C), pH, Incubation Time (h), Agitation (rpm) and Illumination)	Shape, Size (nm) and Location of NPs	References
Gold Nanoparticles			
Stenotrophomonas maltophilia	Cell mass, gold chloride, 25, NA, 8, NA, NA	Spherical, 40, In	(Nangia et al. 2009)
Escherichia coli DH5α	Cell mass, chloroauric acid, 37, NA, 120, shaking, NA	Spherical, triangles and quasi-hexagons, 20, Ex	(Du et al. 2007)
Pseudomonas aeruginosa ATCC 90271, *Pseudomonas aeruginosa*	CFS, hydrogen tetra-chloroaurate, 37, NA, 24, NA, NA	NA, 40, Ex, NA, 25, Ex, NA, 15, Ex	(Husseiny et al. 2007)
Rhodopseudomonas capsulata	Cell mass, chloroauric acid, RT, 4.0 and 7.0, 48, NA, NA	Spherical at pH 7, 10–20, Ex triangular at pH 4, 50–400, Ex	(He et al. 2007)
Rhodococcus sp.	mycelial mass, chloroauric acid, 27, NA, 24, 200, NA	Monodisperse, 9, In	(Ahmad et al. 2003a)
Marinobacter pelagius	Cell mass, chloroauric acid, NA, 5.0–6.0, NA, NA, NA	Monodisperse, spherical and triangular, 10, Ex	(Sharma et al. 2012)
Plectonema boryanum UTEX485	Cell mass, chloroauric acid, 25, NA, 24, NA, NA	octahedral and sub-octahedral shapes, < 20, In	(Lengke et al. 2006)
Streptomycetes strain HM10	mycelial mass, chloroauric acid, 28, NA, NA, 120 rpm, NA	Spherical, 18, In	(Balagurunathan et al. 2011)
Thermomonospora sp.	mycelial mass, AuCl$_4$, 50, NA, 120, 200, dark	Spherical, 8, Ex	(Ahmad et al. 2003b)
Thermomonospora sp.	mycelial mass, gold chloride, 50, 9.0, 120, agitated at 200 rpm, dark	spherical, 8, Ex	(Sastry et al. 2003)
Streptomycetes sp.	mycelial mass, chloroauric acid, 28, NA, 120, agitated at 120 rpm, NA	spherical and rod-shaped gold, 18–20, In	(Balagurunathan et al. 2011)
Calothrix pulvinata ALCP 745A	Cell culture, HAuCl$_4$, NA, NA, 30 d, NA, NA	AuNP-Spherical, 5.4 nm, In and 5.7 nm, Ex	(Brayner et al. 2007)
Anabaena flosaquae ALCP B24	Cell culture, HAuCl$_4$, NA, NA, 20 d, NA, NA	NA, 12, In and 25, Ex	(Brayner et al. 2007)
Leptolyngbya foveolarum ALCP 671B	Cell culture, HAuCl$_4$, NA, NA, > 100, NA, NA	NA, 6, In and 6.2, Ex	(Brayner et al. 2007)

Copper and Zinc based Nanoparticles

Organism	Conditions	Properties	Reference
Morganella morganii RP42	cell culture, 37, 7.0, 24, shaking at 200 rpm, NA	polydispersed, 3–10 nm, Ex	(Ramanathan et al. 2010)
Morganella psychrotolerans	cell culture, 20, 7.0, 24, shaking at 200 rpm, NA	polydispersed, 3–10 nm, Ex	
Enterococcus faecalis	CFS, Copper sulfate, 37, 7.2, 24, agitated at 120 rpm, light	Spherical, 20–90, Ex	(Ashajyothi et al. 2015)
Pseudomonas fluorescens	cell mass and cell free supernatant, Copper sulfate, 30, NA, 48, 150, NA	spherical and hexagonal, 49, Ex	(Shantkriti and Rani 2014)
Serratia sp.	Cell mass, $CuSO_4$, 37, NA, 48, 220, NA	Polydispersed, 10–30, Ex	(Saif Hasan et al. 2008)
Aeromonas sp.	Cell culture, ZnO, 30, NA, 24, 120, NA	Spherical and oval, 57.72, Ex	(Jayaseelan 2012)
Lactobacillus sporogens	Cell culture, $ZnCl_2$, RT, 6, 9, shaker, NA	Hexagonal, 5–15, Ex	(Prasad and Jha 2009)

Cadmium-based nanoparticles

Organism	Conditions	Properties	Reference
Escherichia coli	Cell mass, cadmium chloride and sodium sulphide, RT, NA, 1.5 hr, NA, NA	2–5 nm, In	(Sweeney et al. 1998)
Clostridium thermoaceticum	Cell culture, $CdCl_2$, NA, NA, 24 h, NA, NA,	NA, NA, Ex	(Cunningham and Lundie 1993)
Escherichia coli	Cell culture, cadmium chloride, sodium sulphide, RT, 7.2, NA, NA	Polydispersed, 2–5 nm, In	(Sweeney et al. 1998)

Micscellaneous

Organism	Conditions	Properties	Reference
Shewanella algae ATCC 51181	Cell mass, chloroplatinic acid, 25, 7.0, NA, NA, NA	(PtNPs), NA, discrete 5 nm, In	(Konishi et al. 2007)
Calothrix pulvinata ALCP 745A	Cell culture, H_2PtCl_6, NA, NA, 15 d, NA, NA	(PdNPs), NA, 3.2, In	(Brayner et al. 2007)
	Cell culture, $Pd(NO_3)_2$, NA, NA, 0.5, NA, NA	(PdNPs), NA, 3.5, In	(Brayner et al. 2007)
Bacillus thuringiensis	Culture supernatant, cobalt acetate, 37, NA, 12, shaking, NA	(CoNPs)[5] polydispersed, spherical, oval, 84.81 nm, Ex	(Marimuthu et al. 2013)

[1]CFS: Cell Free Supernatant, [2]NA: Not Available, [3]Ex: Extracellular, [4]In: Intracellular, [5](CoNPs): Cobalt Nanoparticles

AgNPs having potential antibacterial activity. Some other silver nanoparticles produced by bacteria are listed in Table 1.

2.2 Gold Nanoparticles (AuNPs)

Gold is one of the rarest metals on earth and its importance has been known for millennia. In recent years, the synthesis of AuNPs have been the focus of intense interest because of their stability, resistance towards oxidation and biocompatibility, leading to their emerging applications in a number of areas, such as bioimaging, biosensors, biolabels, biomedicines and so forth. Researchers are now concentrating on the biosynthesis of metal nanoparticles using both uni- and multi-cellular organisms (Das et al. 2009, Sharma et al. 2012).

Initial studies uncovered the fact that *Bacillus subtilis* 168 can reduce Au^{3+} ions upon incubation with gold chloride solution under ambient conditions and accumulates octahedral gold NPs (5–25 nm) intracellularly. Efficient synthesis of monodisperse AuNPs have also been reported by extremophilic actinomycetes *Thermomonospora* sp. and an alkalotolerant *Rhodococcus* sp. Notably, the continued growth of the cells, even after the biosynthesis of gold NPs, showed that the metal ions were not toxic to the cells (Pugazhenthiran et al. 2009, Thakkar et al. 2010). *Plectonema boryanum* UTEX 485, a filamentous cyanobacterium, when reacted with a gold–thiosulfate complex at 25–100°C for up to 1 month, followed by incubation at 200°C for 1 day, lead to separation of filaments into their constituent cells and released membrane vesicles after 14 days at 25°C; later, the precipitation of irregular gold particles on the surface of membrane vesicles was observed (Lengke et al. 2011).

Iravani (2014) observed that plant-growth-promoting bacteria, isolated from Philippine soils, extracellularly synthesized gold NPs of 10–100 nm diameter upon incubation with gold (III) chloride trihydrate for 7 days at 28°C. Probably, nitrate reductase was one of the enzymes responsible for the bioreduction of ionic gold. Some other gold nanoparticles produced by bacteria are listed in Table 1.

2.3 Copper and Zinc-based NPs

Till date, research in the field of biosynthesis has mainly been focused on AgNPs [7,699 papers, 59%], Au NPs followed by ZnO [4,640 papers, 36%] and, finally, CuO NPs [690 papers, 5%] (Shobha et al. 2014). The few papers in the literature on the synthesis of copper nanoparticles have reported synthesis of copper nanoparticles in the oxide form. This is probably because copper is well known as an oxidation receptor and the most successful chemical synthesis of metallic copper nanoparticles is either carried out in organic phases or requires elaborate setups for aqueous phase synthesis in order to avoid potential oxidation of Cu^{2+} into its oxide forms. The synthesis of pure metallic copper nanoparticles in aqueous phase is, therefore, still an open challenge for bionanotechnologists (Ramanathan et al. 2010).

Different bacteria belonging to the genus *Serratia, Pseudomonas stutzeri, Morganella morganii RP42, Morganella psychrotolerans, Escherichia coli* and *Streptomyces* sp., were used to synthesize CuNPs (Shobha et al. 2014).

Nanoparticles of semiconductors have also attracted attention as a result of their interesting optical and electronic properties. In particular, nanoparticles of the semiconductor, zinc sulphide (ZnS) and Zinc oxide (ZnO), are the most attractive materials for applications in areas such as IR optical devices and fast optical switching devices, biosensors, nanoelectronics and solar cells (Bai et al. 2006, Jayaseelan 2012). The industrial production of ZnONPs presently reaches 550 tons per year (Vankova et al. 2017); however, the reports on the bacterial synthesis of zinc nanoparticles are scanty.

Natural biofilms of sulfate-reducing bacteria from the family *Desulfobacteriaceae* have been reported to synthesize sphalerite (ZnS) particles. This complex ZnS biomineralization was found to significantly reduce the concentration of Zn below acceptable levels for drinking water, suggesting a cheaper and eco-friendly alternative for water purification (Mandal et al. 2006, Popsecu et al. 2010, Ramanathan et al. 2015). Alternatively, a novel, clean biological transformation reaction using immobilized *Rhodobacter sphaeroides* was developed for the synthesis of ZnS nanoparticles. The size of nanoparticles is determined by the culture time of the *Rhodobacter sphaeroides* and the immobilized beads simultaneously act on separating the ZnS nanoparticles from the *Rhodobacter sphaeroides* (Bai et al. 2006).

Since, ZnO nanoparticles possess antibacterial and antifungal activities at lower concentrations, therefore, the thin coatings of such nanoparticles can be used for the preparation of microbe-resistant articles. Moreover, in comparison to traditional antifungal agents, the use of ZnO nanoparticles as an antifungal agent does not affect the soil fertility (Bai et al. 2006, Jayaseelan 2012). A green, low-cost approach, capable of producing ZnO NPs at room temperature using reproducible bacteria *Aeromonas hydrophila*, is reported by (Jayaseelan 2012). The formed ZnO nanoparticles were hydrophilic in nature, highly stable, disperse uniformly in water, and showed significant antimicrobial activity against *Pseudomonas aeruginosa* and *Aspergillus flavus*.

2.4 Magnetic NPs

Magnetic nanoparticles are advantageous in that they can offer selective attachment to a functional molecule, confer magnetic properties to the target, and allow manipulation and transportation to a desired location through the control of a magnetic field produced by an electromagnet or permanent magnet. Because of their size, magnetic nanoparticles are able to overcome the effect of the gravitational field, magnetic field gradient, and the potential magnetic agglomeration which could result when particles come into contact with one another (Vatta et al. 2006).

Magnetite is a common product of bacterial iron reduction and could be a potential physical indicator of biological activity in geological settings. The magnetotactic bacteria synthesize intracellular magnetic particles, comprising iron oxide, iron sulphides, or both (Li et al. 2011). The magnetotactic bacteria, like *Magnetospirillum magnetotacticum* and *M. gryphiswaldense*, possess the ability to produce intracellular phospholipid membrane-bound ferromagnetic particles composed of magnetite (Fe_3O_4) or greigite (Fe_3S_4); the structures thus formed are termed as magnetosomes (Matsunaga and Takeyama 1998, Mandal et al. 2006, Kannan and Subbalaxmi 2010, Revati and Pandey 2011). *Thermoanaerobacter ethanolicus* (TOR-39), a thermophilic fermentative bacterial strain, forms single-domain tiny magnetic particles (< 12 nm) exhibiting octahedral shapes on the outside of the bacterial cells. Transition metals, such as Co, Cr and Ni, may also be substituted for magnetic crystals biosynthesized in the bacterium by way of electrochemical processes. The mineralization processes are highly controlled by the magnetotactic bacteria, leading to the formation of uniform, species-specific magnetic nanoparticles. Sometimes, the particles are assembled into single or multiple chains and anchored inside the cell, enabling the bacteria to passively orient themselves along geomagnetic field lines (Mandal et al. 2006).

Similarly, dissimilatory iron-reducing bacteria of the genus *Geobacter metallireducens* and *Shewanella putrifaciens* are the most commonly studied microbes that generate crystals of magnetite as by-products during their growth. Sulfate-reducing bacteria (SRB), such as *Archaeoglobus fulgidus* and *Desulfuromonas acetoxidans*, produce particles of magnetic iron. The SRB respire with sulfate, anaerobically releasing H_2S (dissimilatory sulfur reduction) (Kannan and Subbalaxmi 2010, Revati and Pandey 2011). Cells of *Desulfovibrio magneucus* strain RS-I, an anaerobic sulfate-reducing bacterium, form intracellular nanocrystals of magnetite but are only weakly magnetotactic (Posfai et al. 2006).

Moreover, *Thiobacillus thiooxidans, Thiobacillus ferrooxidans* and *Sulfolobus acidocaldarius* can reduce ferric ion to the ferrous state while growing on elemental sulfur as an energy source. *T. thiooxidans* was able to reduce ferric iron at low pH medium aerobically, but the bioreduction of ferric iron using *T. ferrooxidans* was not aerobic because of the rapid bacterial reoxidation of the ferrous iron in the presence of oxygen (Iravani 2014).

Recently, the ability of actinomycetes *Actinobacter* sp. to synthesize magnetic iron sulphide nanoparticles extracellularly under aerobic conditions has been reported. These nanoparticles were synthesized with ferric ions in the presence of exogenous sulfate source and typically encompassed Fe_3S_4 and FeS_2 of 20 nm diameter (Ramanathan et al. 2015). Similarly, *Alteromonas putrefaciens*, *Pelobacter acetylenicus* and *P. venetianus* bacteria possess an ability to couple the oxidation of potential electron donors (such as lactate, pyruvate, hydrogen, propanol, butanol and formate) to the reduction of Fe (III) was (Iravani 2014). Some other magnetic nanoparticles produced by bacteria are listed in Table 1.

2.5 Cadmium Sulphide Nanoparticles (CdSNP)

Along with metal NPs, considerable interest of researchers lies towards the development of procedures for the synthesis of semiconductor NPs, such as cadmium sulphide, for application as quantum-dot fluorescent biomarkers in cell labelling.

Initial reports on the intracellular semiconductor nanoparticle synthesis shows that *E. coli*, when incubated with cadmium chloride ($CdCl_2$) and sodium sulphide (Na_2S), spontaneously formed CdS semiconductor nanocrystals. They showed that the formation of nanocrystals was markedly affected by physiological parameters (Sweeney et al. 1998). Similarly, *Clostridium thermoaceticum* precipitated CdS at the cell surface as well as in the medium when exposed to $CdCl_2$, in the presence of cysteine hydrochloride as a source of sulphide in the growth medium (Popsecu et al. 2010, Thakkar et al. 2010). Additionally, *Klebsiella aerogenes*, upon exposure to the Cd^{2+} ions, resulted in the intracellular formation of CdS nanoparticles (20–200 nm) and the composition of the nanoparticles formed was a strong function of buffered growth medium for the bacteria (Sweeney et al. 1998, Sastry et al. 2003, Kannan and Subbalaxmi 2010). Synthesis and characterization of CdSNPs from the marine cyanobacterium, *Phormidium tenue* NTDM05 has also been reported recently (Singh et al. 2015).

Through a recombinant approach, a peptide reported to be able to bind to CdS, termed as CDS 7, which was expressed in *E. coli* and the engineered bacteria were allowed to interact with CdS precursors in order to induce the formation of fluorescent CdS QDs. A series of parameters, including reactant concentration, incubation time of bacteria and reaction time, were also studied (Mi et al. 2011). Some other CdS nanoparticles produced by bacteria are listed in Table 1.

2.6 Other Metal NPs

Dissimilatory metal and sulfate-reducing bacteria can couple the reduction of dissolved U (VI) to the oxidation of organic matter and H_2, resulting in the precipitation of biogenic UO_2 (bio-UO_2). Biogenic uraninite was produced by *Shewanella oneidensis* MR-1 and *S. putrefaciens*. Uraninite particles were found to associate with and accumulate on the extracellular polymeric substances secreted by *S. oneidensis*. However, the synthesis of biogenic UO_2 mediated by *S. putrefaciens* cell suspensions exhibited the presence of a coating of an organic matter, which inhibited particle annealing and protected it from the surrounding environment (Duran and Seabra 2012).

The palladium NPs (PdNps) were also reported to be synthesized by the sulfate-reducing bacterium, *Desulfovibrio desulfuricans,* and metal ion-reducing bacterium, *S. oneidensis* (Li et al. 2011). A new, biologically-inspired method to produce nanopalladium is the precipitation of Pd on a bacterium, i.e., bio-Pd.

Citrobacter braakii, Shewanella oneidensis MR-1 as well as an autoaggregating mutant (COAG) resulted in precipitation of palladium Pd(0) NPs on the cell wall and inside the periplasmic space (bioPd). This bio-Pd can be applied as a catalyst in dehalogenation reactions (Windt et al. 2005, Hennebel et al. 2011). Resting cells of the metal ion-reducing bacterium *Shewanella algae* were found to accumulate elemental platinum NPs (PtNPs) (5 nm) into in their periplasm, by reducing aqueous $PtCl_6^{2-}$ ions, at room temperature and neutral pH within 60 min with lactate as the electron donor (Konishi et al. 2007).

Similar, biomineralization events, such as the formation of tellurium (Te) in *Escherichia coli* K12, the direct enzymatic reduction of Tc (VII) by resting cells of *Shewanella putrefaciens* and *Geobacter metallireducens*, the reduction of selenite to selenium by *Enterobacter cloacae, Desulfovibrio desulfuricans* and *Rhodospirillum rubrum*, and the formation of crystalline and needle-like nanoparticles of Lanthanum by *Pseudomonas aeruginosa* have also been reported (Iravani 2014). Likewise, silicon/silica nanocomposites were reported to be synthesized by the *Actinobacter* spp. upon exposure to K_2SiF_6 precursor. The reductases and oxidizing enzymes produced under the stimulation of the precursor probably play a key role in the biosynthesis process of silicon/silica nanocomposites (Singh et al. 2008). Uniform-sized monodispersed intracellular mercury NPs (2–5 nm) were also found to be synthesized by *Enterobacter* sp. The culture conditions (pH 8.0 and lower concentration of mercury) promote the synthesis of uniform-sized spherical, and monodispersed intracellular mercury nanoparticles (Li et al. 2011).

Reportedly, tellurite has been reduced to elemental tellurium by two anaerobic bacteria, *Bacillus selenitireducens* and *Sulfurospirillum barnesii*. *B. selenitireducens* initially formed nanorods which later clustered together to form larger rosettes of 1000 nm, while *S. barnesii* formed small irregularly shaped extracellular nanospheres of diameter 50 nm. Similarly, tellurium-transforming *Bacillus* sp. BZ isolated from the Caspian Sea in northern Iran was used for the intracellular synthesis of elemental tellurium NPs (Iravani 2014).

Equally, an anaerobic hyperthermophilic microorganism, *Pyrobaculum islandicum*, reduces many heavy metals, including U (VI), Tc (VII), Cr (VI), Co (III), and Mn (IV), with hydrogen as the electron donor (Li et al. 2011). Additionally, many Fe (III)-reducing microorganisms could reduce forms of oxidized metals, including radio nuclides, such as uranium (VI) and technetium (VII), and trace metals, including arsenic (V), chromium (VI), cobalt (III), manganese (IV) and selenium (VI) (Iravani 2014).

Additional examples of nanostructures in biology include bacteria that play an active role in the precipitation of carbonates or in the formation of calcium pyrophosphate crystals and S-layer-producing bacteria (Klaus-Joerger et al. 2001).

3. Mechanism of NPs Synthesis by Bacteria

Nanoparticles are biosynthesized by bacteria when they capture target ions from their environment and then turn them into element metal with the help of cellular enzymes. The process of biosynthesis of NPs has been classified as intracellular and extracellular synthesis, according to the location of formation of NPs. The extracellular synthesis offers a great advantage over an intracellular process of synthesis from the application point of view. Principally, intracellular synthesis of Nps may achieve a superior control over the size and shape distributions of the NPs, but product harvesting and recovery are more troublesome as well as uneconomic. The extracellular synthesis is comparatively more adaptable to the synthesis of a variety of NPs. Extracellular synthesis of NPs facilitates attainment of pure and monodisperse NPs, that are free from cellular components of organisms and easy downstream processing. The surface trapped nanoparticles or those formed inside the cell would require an additional step of processing, such as ultrasonication, for their release into the surrounding liquid media. Therefore, in recent times, the focus of research has been given to the development of an extracellular process which offers a great advantage over the intra-cellular process of synthesis from the application point of view (Saifuddin et al. 2009).

Different microorganisms have different mechanisms of forming nanoparticles. One of the general mechanisms involved during the extracellular metal synthesis of NPs by microorganisms is bioreduction. The negatively charged cell wall of the bacteria interacts electrostatically with the positively charged metal ions and, therefore, plays a crucial role in the extra as well as intra-cellular synthesis of NPs. This in turn is followed by enzymatic reduction of the metal ions leading to their aggregation and the formation of NPs. This reduction of metal ions is facilitated by specific reducing enzymes like NADH-dependent reductases or nitrate-dependent

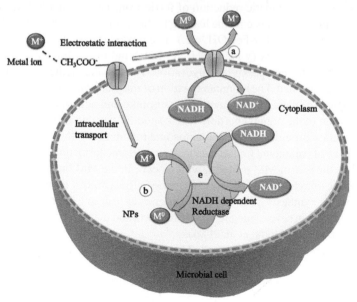

Fig. 1. Schematic representation of the synthesis of metal nanoparticles by bacteria (modified from Sintubin et al. 2009).

reductases. Alternatively, during the intracellular synthesis, the smaller sized NPs may diffuse through the microbial cell wall after getting reduced by the enzymes present within the cell wall (Ramezani et al. 2010, Li et al. 2011, Rai et al. 2011). A hypothetical mechanism applicable to all bacteria is provided in Fig. 1.

Bacteria *R. capsulate and Stenotrophomonas maltophilia,* while bioreducing Au^{3+} ions to Au^0 for the formation of AuNPs, are reported to secrete cofactor NADH- and NADH-dependent enzymes. The reduction seems to be initiated by transfer of electrons from the NADH to NADH-dependent reductase acting as an electron carrier. Finally, the electrons are accepted by the Au ($^{3+}$) ions and are reduced to Au (0). The amino, sulfhydryl and carboxylic groups of the enzymes, secreted by the bacteria, helps the $AuCl^{4-}$ to bind to the bacterial cell and therefore play crucial role in reducing the Au^{3+} ions (He et al. 2007, Nangia et al. 2009).

Li et al. (2011), have illustrated a four steps mechanism for the synthesis of magnetite (Fe_3O_4) NPs. The first step involves the invagination and formation of vesicles in the cytoplasmic membrane, which is probably mediated by a GTPase. The formed vesicles are then assembled, along with the cytoskeletal filaments, into a linear chain. Secondly, the ferrous ions are accumulated into the vesicles by the transmembrane iron transporters. External iron is brought inside the cell by transport proteins and siderophores. Finally, the proteins, which are tightly bound to magnetite, trigger magnetite crystal nucleation and/or regulate morphology. Contrastingly, Duran and Seabra (2012) have illustrated a different four-steps mechanism for the synthesis of magnetite (Fe_3O_4) NPs by *Actinobacter* spp. The first step involves uptake of iron from ferric quinate and its transportation into the cytoplasm via an enzymatic reduction of ferric iron. This is then followed by the partial oxidation of a redox enzyme leading to the formation of low density hydrous ferric oxides (α-FeOOH, γ-FeOOH, etc.). The next step consists of the formation of a ferrihydrite complex (γ-FeO (OH) 2FeOH$^+$) and the addition of ferrous iron to this ferrihydrites complex via dehydration. Finally, magnetite (Fe_3O_4) NP and proton (H^+) are formed. The biomineralization of magnetosome particles is achieved by a complex mechanism that involves the uptake and accumulation of iron and the deposition of the mineral particle with a specific size and morphology within a specific section provided by the magnetosome membrane (MM). Since the MM is thought to be of paramount importance in magnetosome formation, researchers have focused on the role of MM proteins, which occur in the MM but not in the soluble (periplasmic or cytoplasmic) fraction or in the cytoplasmic or outer membranes, in magnetosome synthesis. The mam (MM) genes appear to be conserved in a large gene cluster within several magnetotactic bacteria and may be involved in magnetic biomineralization (Mandal et al. 2006).

The extremophillic actinomycete *Thermomonospora* sp. reduced the gold ions extracellularly, yielding AuNPs. Reportedly, the free amine groups or cysteine residues in the proteins bind to AuNPs and stabilize them. The enzymes that reduces chloroaurate ions and cap the AuNPs formed by the reduction process, may be one of them. Thus, capping and stabilization of the AuNPs is mediated by a different protein (Araya 2010).

Similarly, the mechanism of formation of ZnS nanoparticles by biological transformation reaction of *Rhodobacter sphaeroides* involves entry of soluble sulfate into immobilized beads via diffusion which is then carried to the inner membrane of the cell facilitated by sulfate permease. The sulfate is then reduced to sulphite, by ATP sulfurylase and phosphoadenosine phosphosulfate reductase, which is further reduced to sulphide by sulphite reductase. After this process, sulphide reacts with the soluble zinc salt and the ZnS nanoparticles are synthesized. Finally, ZnS nanoparticles are discharged from immobilized cells to the solution (Bai et al. 2006, Iravani 2014).

4. Bacterial Enzymes Involved in NP Synthesis

A complete knowledge of the molecular mechanisms involved in the microbial synthesis of NPs is necessary in order to control the size, shape and crystallinity of NPs. Therefore, study of enzymes involved in NP biosynthesis is mandatory. The negative electrokinetic potential of the microorganisms enable enzymes to attract the cations and act as trigger of the procedure for the biosynthesis of metal and metal oxide NPs (Ramezani et al. 2010, Raliya and Tarafdar 2013).

4.1 Oxidoreductases

The oxidoreductases are pH sensitive enzymes and work in an alternative manner. At a lower value of pH, oxidase gets activated while a higher pH value activates the reductases. Minaeian et al. (2008), for the first time reported the extracellular synthesis of NPs by a *Klebsiella pneumoniae*, *Escherichia coli* and *Enterobacter cloacae*. They postulated the probable role of reductase/electron shuttle for AgNP synthesis. Similarly, it is proposed that the synthesis of Si/SiO$_2$ nanocomposites by *Actinobacter* sp. and TiO$_2$ NPs synthesis by Lactobacillus strain is mediated by certain non-specific reductases and oxidizing enzymes (Duran and Seabra 2012, Shi et al. 2015).

Nangia et al. (2009) reported that the biosynthesis of AuNPs and their stabilization via charge capping in *Stenotrophomonas maltophilia* involved NADPH-dependent reductase enzyme. The bacterial cells, when incubated with varying concentrations of NADPH (0.05–0.8 mM), caused a colour change in the solution, proving presence of specific reductases in the microorganism, which may be induced by the specific ions. The bacteria *R. capsulate* is known to secrete cofactor NADH- and NADH-dependent enzymes. The bioreduction of gold ions was found to be initiated by the electron transfer from NADH by NADH-dependent reductase as electron (Duran et al. 2011, He et al. 2007, Iravani 2014).

The role of pH-dependent membrane-localized oxidoreductases has also been hypothesized for the synthesis of ZnO nanoparticles in a study using *Lactobacillus sporogenes* (Jain et al. 2013). The synthesis of ZnO NPs by *Aeromonas hydrophila* might have resulted due to variation in the level of rH^2 or pH, which leads to activation of pH sensitive oxidoreductases (Jayaseelan et al. 2012).

In an another study, stable, predominantly monodispersed, and spherical selenium NPs (21 nm) were synthesized using the bacterial isolate *Pseudomonas aeruginosa* strain JS-11. Suggestively, the metabolite phenazine-1-carboxylic acid released by strain JS-11 in culture supernatant, along with the known redox agents like NADH and NADH dependent reductases, was responsible for biomimetic reduction of SeO_3^{2-} to Se^0 nanospheres (Iravani 2014).

4.2 Nitrate/Nitrite Reductase

Another important enzyme that is responsible for NP reduction in some microorganisms is nitrate-dependent reductase. This enzyme, which is usually conjugated with an electron donor (quinine), reduces the metal ion and changes it to elemental form.

Nitrate reductase enzyme from *Bacillus licheniformis* and *Lactobacillus* sp. has been reported to be involved in the synthesis of AgNPs (Dhoondia and Chakraborty 2012, Wei et al. 2012). Similarly, silver ions accumulated on the cell surface of *P. putida* NCIM 2650, grown in the presence of silver nitrate, were further transformed to silver nanoparticles, by the extracellular nitrate reductase enzyme and its activity was measured spectrometrically as 0.958 U/ml/min, optimum at 20 min (Thamilselvi and Radha 2013).

The nitrate reductase enzymes, together with electron shuttling compounds and other peptides/proteins, may be responsible for the reduction of silver ions and the subsequent formation of silver nanoparticles (Duran et al. 2011, Saifuddin et al. 2009).

4.3 Sulfate and Sulphite Reductase

The role of these reductases has been elucidated during the formation of ZnSNPs by *Rhodobacter sphaeroides*. Initially, a soluble sulfate is carried to the interior membrane of *R. sphaeroides* cell facilitated by the enzyme sulfate permease. The sulfate is then subsequently reduced to sulphite by the enzyme ATP sulfurylase and phosphoadenosine phosphosulfate reductase. The next step in the sequence is the reduction of sulphite to sulphide by the enzyme sulphite reductase. The sulphide reacts with O-acetyl serine in order to synthesize cysteine via O-acetylserine thiolyase, and then cysteine produces S^{2-} by a cysteine desulfhydrase in the presence of zinc. After this process, S^{2-} reacts with the soluble zinc salt and the ZnS NPs are synthesized (Bai et al. 2006, Iravani 2014).

The important role of bacterial sulfate reductases for the conversion of sulfate into sulphide during the synthesis of magnetic iron sulphide nanoparticles by *Actinobacter* sp. was postulated, along with the involvement of several low molecular weight proteins in the stabilization of these particles (Ramanathan et al. 2015).

4.4 Glutathione

Glutathiones (GSH), having c-Glu-Cys-Gly as their general structure, are crucial tripeptides involved in several metabolic processes in bacteria, yeasts, plants and animals. Metallothioneins and phytochelatins are the two broad categories of metal-binding ligands, into which GSHs are classified. They are considered as excellent detoxificators, because of their unique redox and nucleophilic properties and are, therefore, actively involved in the bioreduction and defence against free radicals and xenobiotics (Ramezani et al. 2010).

The *Lactobacillus*-mediated biosynthesis of silver nanoparticles (10–25 nm) from yoghurt and probiotic tablets involved metabolically produced pyruvate and lactate, along with the participation of the glutathione and thioredoxins (Duran et al. 2011). In *E. coli*, cellular thiol content in general and glutathione content in particular might be responsible for the observed growth phase dependence of nanocrystal formation. Glutathione and cysteine, thiolates that are able to form high-affinity metal ligand clusters, have also been shown to promote the formation of CdS and ZnS nanocrystals (Sweeney et al. 1998).

4.5 Cysteine Desulfhydrase

The cysteine desulfhydrase enzyme of *Rhodopseudomonas palustris* belongs to the family of lyases, specifically the class of carbon-sulfur lyases. This enzyme is located in cytoplasm of *R. palustris* and the content of cysteine desulfhydrase, depending upon the growth phase of cells, was responsible for the formation of CdS nanocrystal, while protein secreted by the *R. palustris* stabilized the CdSNP. In addition, R. palustris was able to efficiently transport CdS NPs out of the cell (Ramezani et al. 2010, Shi et al. 2015).

The activity of cysteine desulfhydrase was found to depend on different culture time during synthesis of PbS nanoparticles by *R. sphaeroides*, and maximum activity of 46.2 Ug^{-1} was observed after 40 h of incubation (Bai and Zhang 2009). Similarly, comparable levels of cysteine desulfhydrase activity were observed in the extracts of cadmium-treated *Clostridium thermoaceticum*, and the amount of protein extracted and total sulphide was two-fold and four-fold higher in cultures containing cadmium than in control cultures (Cunningham and Lundie 1993).

4.6 Hydrolases

Bhatta et al. (2008) deduced the expression of four different proteins during the synthesis of Si/SiO_2 Nps by *Actinobacter* spp. The molecular weights of these four proteins were 5, 30, 35 and 105 kDa. The enzymes having molecular weights 30 and 35 kDa exhibit reductase activity and could reduce gold or silver ions to their respective metallic NPs. The high molecular weight (105 kDa) protein is postulated to be a hydrolase, involved in hydrolysis of Si ions. These enzymes reduce the stress by converting the toxic Si ions to NPs and thereby precipitating this Si/SiO_2 from solution. Similarly, the formation of NPs of zirconia, silica and titania by fungi is reported to be mediated by certain cationic proteins (21–24 kDa) capable of extracellularly hydrolyzing compounds (Ramezani et al. 2010).

5. Factors Affecting Bacterial Synthesis of NPs

A control over the environmental parameters for nanoparticle synthesis can reportedly lead to the synthesis of NPs of various sizes and shapes. In addition to biomechanical factors, a number of physicochemical factors have an impact on the bioreduction of metal ions into metallic NPs in biological systems (Lengke et al. 2011).

5.1 Effect of Temperature

Different temperature conditions signify different kinetic as well as structural characteristics of Np synthesis. Mild temperature conditions aid in the formation of stable and increased quantities of production with defined dimensional size distribution. At the same time, the higher temperatures lead to increased reaction rates and decreased particle size, as well as narrower size distribution (Kannan and Subbalaxmi 2010).

Similarly, it has been observed that gold nanotriangles formation is kinetically controlled and is highly favoured at the low temperature range. It was demonstrated that temperature plays an important role in controlling the proportion and relative amounts of gold nanotriangles and nanospheres (Araya 2010). Correspondingly, AgNPs of various shapes can be attained by controlling the growth kinetics of *Morganella psychrotolerans*, which were grown at four different temperatures. At the optimum growth temperature (20°C), spherical AgNPs of sizes ranging between 2–5 nm were obtained. On the contrary, at temperatures of 25°C and 15°C, a mixture of triangular and hexagonal nanoplates, along with spherical nanoparticles, were obtained. Nevertheless, the number of nanoplates increased significantly at 4°C, whereas only a small proportion of spherical nanoparticles were obtained (Shi et al. 2015). Similarly, AgNP synthesis by *E. coli* was observed to be heavily influenced by the incubation temperature. AgNPs of 50 nm and 15 nm were synthesized at room temperature and at 60°C, respectively (Lengke et al. 2011).

5.2 Effect of pH

The size of NPs is also strongly influenced by the pH of the medium (Araya 2010). Reportedly, Fe (III) reduction typically results in an increase in pH, ionic strength of the pore water and the concentration of a variety of cations. Moreover, magnetite formation is favoured by high pH (Revati and Pandey 2011).

The synthesis of AgNPs is faster under alkaline conditions as compared to acidic conditions. Synthesis of AgNPs increases as the pH increases towards the alkaline region and reaches the maximum at pH 10, beyond which the rate of the NP synthesis decreases. Additionally, the need to agitate the mixture for the formation of AgNPs is curtiled under alaline conditions and all the silver ions supplied will be converted to AgNPs in just 30 min in contrast to the requirement of 4 days for the production of silver ions in normal conditions. Moreover, under alkaline conditions, the ability of the enzyme involved in the synthesis of AgNP increases. The size-controlled synthesis of AgNP by controlling the environment is because of the formation of many seed crystals. At acidic pH and lower temperatures there will be less nucleation for the formation of silver crystals, on which new incoming silver atoms deposit to form larger sized particles. The dynamics of the ions increase with the increase in the pH and temperature, in addition, more nucleation regions are formed due to the availability of $-OH$ ions and increased temperature. The conversion of Ag^+ to Ag^0 increased in response to an increase in the kinetics of the deposition of the silver atoms. Similarly, AgNPs of size 45 nm were formed by *E. coli* at acidic pH, whereas at pH 10 the size was just 15 nm (Lengke et al. 2011).

Lactobacillus casei is an acidophilic bacterium and has better growth and metabolism in a weak acidic environment; therefore, reduction in the pH of the reaction mixture leads to an increase in the production of colloidal AgNPs by the bacterium (Korbekandi et al. 2012).

5.3 Effect of Microbial Species

Iron oxide magnetic particles of uniform sizes and morphologies are generated by magnetotactic bacteria. Cuboid, bullet-shaped, rhombic and rectangular are the various shapes acquired by the magnetites formed by magnetotactic bacteria. Crystal morphologies and compositions of various types have been witnessed, that are species or strain dependent, indicating the presence of stringent biological control (Li et al. 2011).

Cell-free culture supernatants of five psychrophilic bacteria, *Phaeocystis antarctica*, *Pseudomonas proteolytica*, *Pseudomonas meridiana*, *Arthrobacter kerguelensis* and *Arthrobacter gangotriensis*, and two mesophilic bacteria, namely *Bacillus indicus* and *Bacillus cecembensis*, have been used to biosynthesize silver NPs. It was observed that the supernatant of *A. kerguelensis* could not produce AgNPs at the temperature, whereas *P. antarctica* could synthesize silver NPs. Therefore, this study provided important evidence that the factors in the cell-free culture supernatants which facilitated the synthesis of silver NPs varied between different bacterial species (Iravani 2014).

5.4 Effect of Radiations

Visible light has been shown to increase the biosynthetic rate of AgNPs by bacteria. Solar radiation was applied to AgNPs synthesis by cell-free extracts of *Bacillus amyloliquefaciens* LSSE-62. Promisingly, the use of natural sunlight would also reduce energy costs (Wei et al. 2012).

The kinetics of silver nanoparticles synthesis using the cell filtrates in combination with microwave irradiation indicated that the rapid synthesis of nanoparticles would be suitable for developing a "green nanotechnology" biosynthesis process for mass scale production. The advantage of using microwave radiation is that it provides uniform heating around the nanoparticles and can assist the digestive ripening of NPs without aggregation (Saifuddin et al. 2009).

6. Applications of NPs in Agriculture

Indiscriminate use of pesticides and fertilizers leads to environmental pollution, emergence of agricultural pests and pathogens, and loss of biodiversity. Nanotechnology, by virtue of nanomaterial-associated properties, has potential agro-biotechnological applications for alleviation of these problems (Ghormade et al. 2011).

Nanotechnology offers efficient control and precise release of pesticides, herbicides and fertilizers. Development of nanosensors can help in determining the required amount of farm inputs, such as fertilizers and pesticides. Nanosensors for detection of pesticide residues offer high sensitivity, low detection limits, fast responses, super selectivity and small sizes. They can also detect the levels of soil moisture and soil nutrients (Sabir et al. 2014). The details of some of the applications of nanoparticles are discussed here:

6.1 Growth Promotion

Recent researches on NPs in a number of crops, including corn, wheat, ryegrass, alfalfa, soybean, tomato, radish, lettuce, spinach, onion, pumpkin, bitter melon and cucumber, have provided evidence of enhanced seedling growth, germination, photosynthetic activity, nitrogen metabolism, protein level, mRNA expression and changes in gene expression, indicating their potential use for crop improvement (Shobha et al. 2014). An account of the positive impact of the treatment with different metal nanoparticles on the growth of different plant varieties has been depicted in Table 2.

Reportedly, magnetic nanoparticles coated with tetramethylammonium hydroxide led to an increase in chlorophyll-a level in maize. Similarly, use of iron oxide in pumpkin was also observed to increase root elongation that was attributed to the iron dissolution (Sekhon 2014).

Reportedly, spinach seeds treated with Nano-TiO_2 had increased growth rate as compared to untreated seeds. The key reason for the increased growth rate could have been the photo-sensitization and photo-generation of "active oxygen-like superoxide and hydroxide anions" by nano-TiO_2 that can escalate the seed stress resistance as well as promote capsule penetration for water and oxygen intake

Table 2. Growth promoting effect of different nanoparticles.

NP, Concentration applied	Plant Variety, Part, Age	Growth Promoting Impact (% increase in character studied[†])	References
ZnO, 20 mg/L	mung bean (*Vigna radiate*), seeds, NA	root length (42.03) shoot length (97.87) root biomass (40.89) shoot biomass (76.04)	(Mahajan et al. 2011)
ZnO, 1 mg/L	gram (*Cicer arietinum*)	Root length (53.13) shoot length (6.38) root biomass (37.15) shoot biomass (26.61)	(Mahajan et al. 2011)
ZnO, 1 mg/L	tissue culture grown shoots of *Stevia rebaudiana* Bertoni, NA	% of shoot formation (89.6) Mean shoot length (11.11) Mean no. of leaves (38.0) Fresh weight of shoots (55.55) steviol glycosides (rebaudioside A and stevioside) production (NA)[§] antioxidant activities (NA)[§]	(Javed et al. 2017)
ZnO, 10 mg/L	clusterbean (*Cyamopsis tetragonoloba* L.), foliar, 14 d	shoot length (31.5) root length (66.3) root area (73.5) chlorophyll content (276.2) total soluble leaf protein (27.1) rhizospheric microbial population (11–14) acid phosphatase activity (73.5%), alkaline phosphatase activity (48.7) phytase activity (72.4) plant biomass (27.1) gum content (7.5)	(Raliya and Tarafdar 2013)

Table 2 contd....

...Table 2 contd.

NP, Concentartion applied	Plant Variety, part, age	Growth promoting impact (% increase in character studied*)	Reference
ZnO, 10 mg/L	pearl millet cv. HHB 67 (*Pennisetum americanum* L.), Foliar	shoot length (15.1) root length (4.2) root area (24.2) chlorophyll (24.4) total soluble leaf protein (38.7) plant dry biomass (12.5) grain yield (37.7) plant zinc concentration (10.4)	(Tarafdar et al. 2014)
ZnO, various	tomato (*Solanum lycopersicum* L.), foliar and soil	germination percentage (51.8)[2] shoot (10.6)[3] root length (NS[4])[3] Fruit yield by weight (81.9 and 305.4***) Lycopene content (113.1)[5] Dry biomass yield (40.7)[6] Chlorophyll (62.67–227.42)[7]	(Raliya et al. 2015b)
TiO₂, various	tomato (*Solanum lycopersicum* L.), foliar and soil	germination percentage (72.5)[2] Lycopene content (80.2)[5] Dry biomass yield (69.6)[6]	(Raliya et al. 2015b)
TiO₂, 10 mg/L	mung bean (*Vigna radiata*), Foliar, 42 days	shoot length (17.02) root length (49.6) root area (43.0) root nodule (67.5) chlorophyll content (46.4) total soluble leaf protein (94.0) rhizosphere microbial population (48.1) acid phosphatase activity (67.3) alkaline phosphatase activity (72.0) phytase activity (64.0) dehydrogenase activity (108.7)	(Raliya et al. 2015a)
TiO₂, NA	Spinach, seeds, 30 d old	dry weight (73) chlorophyll-a formation (45) photosynthetic rate (300)	(Khot et al. 2012)

Cu-chitosan, 0.04% (w/v)	Maize, seedling, 10 d	Germination % (≈ 60.0) Shoot length (≈ 73.3) Seedling length (≈ 200.0) Fresh weight (≈ 200.0) Dry weight (≈ 300.0) Seed vigor index length (≈ 240.0)	(Saharan et al. 2016)
nano-ZnO, 20 mg/L	mung (*Vigna radiate*, seedlings, foliar, 12 d)	Shoot length (6.47) Wet root biomass (31.79) Wet shoot biomass (38.36) Dry root biomass (39.59) Dry shoot biomass (44.09)	(Dhoke et al. 2013)
nano-FeO, 50 mg/L	shoot growth of mung (*Vigna radiata*) seedlings	Shoot length (10.25) Wet root biomass (30.15) Wet shoot biomass (50.44) dry root biomass (68.16) dry shoot biomass (47.61)	(Dhoke et al. 2013)
nano-ZnCuFe-oxide, 50 mg/L	shoot growth of mung (*Vigna radiata*) seedlings	Shoot length (15.71) Wet root biomass (58.79) Wet shoot biomass (85.76) dry root biomass (42.45) dry shoot biomass (83.92)	(Dhoke et al. 2013)

[1] as compared to control.
[2] 1000 mg/Kg of NP and plant harvested on 5th day.
[3] 250 mg/Kg of NP and plant harvested on 28th day.
[4] NS = not significant. *** results for foliar and soil treatment respectively.
[5] 100 mg/kg of NP by foliar application.
[6] 250 mg/kg of NP by aerosol exposure and harvested after 28 days.
[7] 500 mg/Kg of NP by soil and foliar application respectively.
[8] NA: Not Available.

needed for fast germination. The authors concurred that the nano size of TiO_2 might have increased the absorption of inorganic nutrients, accelerated the breakdown of organic substances and also caused quenching of oxygen-free radicals formed during the photosynthetic process, hence increasing the photosynthetic rate (Khot et al. 2012).

Summarily, once the nanoparticles are up taken by the plant, they are bio-distributed through the entire plant by the vascular network. However, the accumulation rate in tissue is different for foliar and soil application. TiO_2 nanoparticles increase the light absorption and chlorophyll content in the plant. Zinc oxide nanoparticles had a twin role of being an essential nutrient and a co-factor for nutrient mobilizing enzymes. Selecting the proper concentration of nanoparticles is important for realizing higher benefits for a target agro-economic trait (Raliya et al. 2015b).

NPs are promising as efficient nutrient source for plants to increase biomass production due to enhanced metabolic activities and utilization of native nutrients by promoting microbial activities. As food demand is increasing day by day, the yield of staple food crops is much too low. So, it is the need of the hour to commercialize metal nanoparticles for sustainable agriculture.

6.2 Nanopesticides

Evidence indicates that pests cause a loss of 25% in rice, 5–10% in wheat, 30% in pulses, 35% in oilseeds, 20% in sugarcane and 50% in cotton. Therefore, for developing countries like India, the crop yield can be maximized by developing drought- and pest-resistant crops (Dimetry and Hussein 2016). The regulatory bodies, like the Central Insecticide Board (CIB, India) and the Food and Drug Administration (FDA, USA), have increased the stringency of the guidelines for the registration of pesticides. Therefore, there is an urgent need to impart a benign safety profile and improved biological efficacy to the existing chemical and biological pesticides with possible use of delivery systems. Additionally, excessive use of pesticides is deleterious to the environment and causes water pollution. A possible solution to this problem is to enhance the retention time of pesticides (Ali et al. 2014).

Nanopesticides involve either very small particles of active pesticidal ingredients or other small engineered structures with useful pesticidal properties. Nanopesticides can increase the dispersion and wettability of agricultural formulations (i.e., reduce organic solvent runoff), and unwanted pesticide movement. Nanomaterials and biocomposites exhibit the useful properties, such as stiffness, permeability, crystallinity, thermal stability, solubility, and biodegradability, needed for formulating nanopesticides. Nanopesticides also offer large specific surface area and, therefore, increased affinity to the target (Khot et al. 2012). Technologies such as encapsulation and controlled release methods have revolutionized the applications of pesticides and herbicides. Pesticides inside NPs are being designed so that they can be timely released or have release linked to an

Table 3. Nanocidal effect of metal nanoparticles.

NP	Microorganism/Pest	Nanocidal Effect	References
AgNP	Damage to sclerotium forming phytopathogenic fungi	Separation of layers of hyphal wall and collapse of hyphae	(Ali et al. 2014)
AgNP	Unidentified fungal species of the genus Raffaelea causing mortality of oak trees	Harmful effects of Ag NPs on conidial germination	(Ali et al. 2014)
AgNP, ZnO and TiO$_2$	rice weevil caused by Sitophilus oryzae	95% mortality, NA, NA	(Dimetry and Hussein 2016)
silica–silver nanoparticles	Control of plant pathogenic fungi viz., *Botrytis cinerea*, *Rhizoctonia solani, Colletotrichum gloeosporioides, Magnaporthe grisea* and *Pythium ultimum*	NA	(Dimetry and Hussein 2016)
silica–silver nanoparticles	Powdery mildew disease of pumpkin	Pathogens were disappeared from the infected leaves within 3 days of spraying	(Dimetry and Hussein 2016)
ZnO	Inhibited growth of fungus Botrytis cinerea	Deformation in mycelia mats	(Agrawal and Rathore 2014)
ZnO	Penicillium expansum	Inhibited the growth of conidiophores and conidia of	(Agrawal and Rathore 2014)

environmental trigger, resulting in enhanced production of crops and minimum injury to agricultural workers (Ditta 2012, Prasad et al. 2014, Agrawal and Rathore 2014). Nanopesticides can increase the dispersion and wettability of agricultural formulations, thereby reducing their wash off and unwanted pesticide movement. Nanopesticides also offer large specific surface area and, therefore, increased affinity to the target (Khot et al. 2012). An account of the effects of nanoparticle based nanocides is listed in Table 3.

Encapsulation of pesticidal nanoparticles allows for proper absorption of the chemicals into the plants. Many companies make formulations which contain nanoparticles within the size ranges of 100–250 nm; they are able to dissolve in water more effectively than existing ones (thus increasing their activity). Other companies employ suspensions of nanoscale particles (nanoemulsions), which can be either water or oil-based and contain uniform suspensions of pesticidal or herbicidal nanoparticles in the range of 200–400 nm (Prasad et al. 2014).

Crop loss to the turn of 30% in plants is usually caused due to the insect pests infesting several crop plants. The use of chemical insecticides and pesticides for crop protection affects the soil health, water bodies and consequently affects human health. However, very few findings were reported for the insect pest management. It needs more attention for crop protection, to meet the satisfactory level of crop production and to increase economic status of our country (Singh et al. 2015).

6.3 Nano-Microbicide

Silver nanoparticles possess strong bactericidal and broad-spectrum antimicrobial properties and also act against various plant diseases caused by spore-producing fungal pathogens. The effectiveness of AgNPs can further be improved by applying them well before the penetration and colonization of fungal spores within the plant tissues. Silver ions are very reactive and are known to bind with various vital components (DNA and proteins); inhibit enzymes such as NADH dehydrogenase II in the respiratory system; induce reactive oxygen species leading to the formation of highly reactive radicals that destroy the cells (Rai and Ingle 2012, Shi et al. 2015).

The *in vitro* and *in vivo* evaluations of the antifungal action of both silver ions and nanoparticles on *Bipolaris sorokiniana* and *Magnaporthe grisea* showed decreased disease development by phytopathogenic fungi (Agrawal and Rathore 2014, Dimetry and Hussein 2016).

Copper nanoparticles in soda lime glass powder showed efficient antimicrobial activity against gram-positive and gram-negative bacteria, as well as fungi (Dimetry and Hussein 2016).

6.4 Nano Herbicide

Conventional herbicides sprayed to kill weeds can unintentionally also affect the crops and cause huge losses in the crop yield. On the contrary, nanoherbicides mingle with the soil particles and try to destroy the entire weed infestation from the roots up, without affecting the good crops. Target specificity of the nanoherbicides enables them to kill the weeds in order to get better yield. Nanoencapsulation of widely-used herbicides, such as triazine, ametryn and atrazine, when applied leads to an 84% more efficient release to the plants. Continuous use of chemical herbicides makes soil lose all its nutrients and become resistant to the plants, therefore, application of modified AgNPs and carboxy methyl cellulose is required to stimulate the degradation of herbicides (Rameshaiah et al. 2015).

6.5 Nanosensing

Wireless sensor network for precision agriculture where real time data of the climatological and other environmental properties are sensed and relayed to a central repository. Since many of the conditions that a farmer may want to monitor (e.g., the presence of plant viruses or the level of soil nutrients) operate at the nano-scale, and because surfaces can be altered at the nano-scale to bind selectively with particular biological proteins, sensors with nano-scale sensitivity will be particularly important in realizing the vision of smart fields. Precision Agriculture entails the monitoring of various parameters which depend on the type of the crop being harvested (Khot et al. 2012). The vision of smart fields involves sensors for the identification and monitoring of pests, drought or increased moisture levels in order to counterbalance their adverse effects on crop production (Ditta 2012).

Intuitively, nanoparticles can be used as biomarkers or as a rapid diagnostic tool for detection of bacterial, viral and fungal plant pathogens. Researchers have used nano-gold based immuno-sensors that could detect Karnal bunt (*Tilletia indica*) disease in wheat using surface plasmon resonance (SPR). Additionally, plants respond to different stress conditions through physiological changes such as induction of systemic defence, probably regulated by plant hormones: Jasmonic acid, methyl jasmonate and salicylic acid. This indirect stimulus was successfully harnessed in order to develop a sensitive electrochemical sensor, using modified gold electrode with copper nanoparticles, to monitor the levels of salicylic acid in the oil seeds to detect the fungi (*Sclerotinia sclerotiorum*). Researches on similar sensors and sensing techniques needs to be expanded for detecting pathogens, their by-products, or monitor physiological changes in plants and then apply pesticides and fertilizers as needed prior to the onset of symptoms (Khot et al. 2012). A mixture of titanium dioxide, aluminium and silica was reported to effectively control downy and powdery mildew of grapes, probably through direct action on the hyphae, interference with fungal mechanism of recognition of plant surface and stimulation of plant physiological defences. A new composition of nano-silver combined with silica molecules and water-soluble polymer proved effective in suppressing the growth of many plant pathogenic fungi and bacteria. *Pythium ultimum*, *Magnaporthe grisea*, *Colletotrichum gloeosporioides*, *Botrytis cinere* and *Rhyzoctonia solani* showed 100% inhibition of growth at 10 ppm concentration; whilst, *Bacillus subtilis*, *Azotobacter chrococuum*, *Rhizobium tropici*, *Pseudomonas syringae*, *Xanthomonas compestris* pv. and *Vesicatoria* showed 100% growth inhibition at 100 ppm concentration of the nanosized silica-silver (Sharon et al. 2010, Mishra and Singh 2014).

Nanosensors for pesticide residue such as methyl parathion, parathion, fenitrothion, pirimicarb, dichlorvos and paraoxon detection offer, high sensitivity, low detection limits, super selectivity, fast responses and small sizes (Khot et al. 2012). Nanoparticles are 'mini-laboratories' that have the potential to precisely monitor temporal and seasonal changes in soil-plant system. Nanosensors detect the availability of nutrients and water precisely, which is very much essential to achieve the mission of precision agriculture (Chinnamuthu and Boopathi 2009).

6.6 Nano Fertilizers

Substituting nanofertilizers for traditional methods of fertilizer application is a way of releasing nutrients into the soil gradually and in a controlled manner, thus preventing eutrophication and pollution of water resources (Sekhon 2014). Today, increasing micronutrient concentrations of stable food crops, especially in cereal grains, represents an important humanitarian challenge and a high-priority research area. Soil and foliar application of micronutrient fertilizer can be used for several different mineral micronutrients to varying degrees of effectiveness. Agronomic biofortification, especially in the case of foliar application, is highly effective for zinc and selenium, while also being effective for iodine and cobalt (Cakmak

2008). Nanostructured fertilizers can increase the nutrient use efficiency through mechanisms such as targeted delivery and slow or controlled release.

Phosphorus (P) is a limiting factor in plant growth and productivity in almost half of the world's arable soil, and its uptake in plants is often constrained because of its low solubility in the soil. To avoid repeated and large quantity application of rock phosphate as a P fertilizer and enhance the availability of native P acquisition by the plant root surface, the first holistic study focusing on native P mobilization using biosynthesized ZnO nanoparticles in the life cycle of mung bean plants has been reported. Zinc acts as a cofactor for P-solubilizing enzymes, e.g., phosphatase and phytase, and nano-ZnO increased their activity. Analytical studies results showed that ZnO nanoparticles were distributed in all plant parts, including seeds. However, the concentration of Zn was within the limit of the dietary recommendation (Raliya et al. 2016). Similarly, enhanced activity of the enzymes, alkaline phosphatase, acid phosphatase and phytase, was noticed when ZnNPs were used as nanofertilizer to enhance crop production in pearl millet (*Pennisetum americanum* L.) cv. HHB 67 (Tarafdar et al. 2014).

Similarly, treatment with TiO_2 nanoparticles in maize had a considerable effect on growth, whereas the effect of TiO_2 bulk treatment was negligible. Titanium nanoparticles increased light absorption and photo energy transmission. Likewise, a compound of SiO_2 and TiO_2 nanoparticles increased the activity of nitrate reductase in soybeans and intensified plant absorption capacity, making its use of water and fertilizer more efficient. Similarly, use of iron oxide in pumpkin was also observed to increase root elongation that was attributed to the iron dissolution (Sekhon 2014).

Nanofertilizer is a plant nutrient which is more than a fertilizer because it not only supplies nutrients for the plant but also revives the soil to an organic state without the harmful factors of chemical fertilizer. One of the advantages of nanofertilizers is that they can be used in very small amounts. An adult tree requires only 40–50 kg of nanofertilizer while an amount of 150 kg would be required in the case of ordinary fertilizers. Nanopowders can be successfully used as fertilizers and pesticides as well. The yield of wheat plants, grown from seeds treated with metal nanoparticles, increased by 20–25%, approximately (Sabir et al. 2014). This invention has opened new doors for fertilizer industries to produce 'Bionanofertilizer' for enhancing plant nutrition.

6.7 Soil and Water Clean-up

One of the processes using nanoparticles is photocatalysis. This chemical reaction is promoted by light and enhanced by the presence of a catalyst, in this case a nanocatalyst. Nanocatalysts are excellent oxidizing agents and include metal oxides, like TiO_2, ZnO, etc., as well as sulphides, like ZnS. This process can be used for the decomposition of pathogens, heavy metals and many toxic compounds, such as pesticides, which take a long time to degrade under normal conditions (Ya et al. 2015).

A nano-clean up method of injecting nano-scale iron into a contaminated site is developed. The particles flow along with the ground water and decontaminate in route, which is much less expensive than digging out the soil to treat it. This nano scale iron remained active in the soil for 6–8 weeks, after which, it dissolved in the ground water and became indistinguishable from naturally occurring iron (Jha et al. 2011). Similarly, nanocrystals, such as monodisperse magnetite (Fe_3O_4), have a strong and irreversible interaction with arsenic while retaining their magnetic properties. A simple handheld magnet can be used to remove nanocrystals and arsenic from water. Such a treatment could be used for an irrigation water filtration process (Prasad et al. 2014).

7. Issues and Future Scope

The use of micro-organisms in the deliberate synthesis of NPs is a relatively new and promising area of applied research. The microbially synthesized NPs have displayed a wide range of potential applications in different sectors of agriculture and environment. However, despite its extensive applications, bacterial synthesis of NPs still has some challenges to overcome:

- Better control over NP size and shape.
- The scalability of bacterial NP synthesis.
- The establishment of low-cost recovery techniques.
- Elucidation of the mechanistic aspect of nanoparticle biosynthesis by these organisms.
- Understanding the surface chemistry of biosynthesized nanoparticles.
- Use of engineered organisms which can over-express the proteins involved in NP synthesis.
- The toxicity of NPs to humans.
- Sustained financial investment in nanotechnology research and development, technology transfer models, good governance and intellectual properties.
- Efforts to understand and facilitate technology adoption and sharing among industrialized and technologically disadvantage countries (Sastry et al. 2003, Khot et al. 2012, Sharma et al. 2012).

8. Summary and Conclusion

The synthesis of metallic nanoparticles is a dynamic area of academics and, more importantly, "application research" in nanotechnology. However, the biological systems have been relatively unexplored, and there are many opportunities for budding nanobiotechnologists to use the biological systems for metallic nanoparticle synthesis. Importantly, for commercialization purposes, it would be advantageous to have a non-pathogenic biological system that produces the metallic nanoparticles. Elucidation of the detailed mechanism behind biosynthesis of NPs using bacterial systems can lead to imperative modifications in the protocol for augmentation in the overall yield of NPs. Genetic engineering implications to the present process can be achieved if the involved enzymes/proteins and their respective genes can be identified, making bacterial synthesis comparable to commercially-used physical and chemical methods of nanoparticle synthesis. Nanofertilizers can slowly release the required mineral nutrient in response to the immediate environment and demand in the plant systems. However, the ethical and safety issues surrounding the use of nanofertilizers in plant productivity are limitless and must be very carefully evaluated before adapting the use of so called 'nanofertilizers' in agricultural fields.

References

Agrawal, S. and P. Rathore. 2014. Nanotechnology pros and cons to agriculture: A review. Int. J. Curr. Microbiol. App. Sci. 3(3): 43–55. doi:10.13140/2.1.1648.1926.

Ahmad, A., S. Senapati, M.I. Khan, R. Kumar, R. Ramani, V. Srinivas and M. Sastry. 2003a. Intracellular synthesis of gold nanoparticles by a novel alkalotolerant actinomycete, *Rhodococcus* species. Nanotechnology 14(7): 824–828. doi:10.1088/0957-4484/14/7/323.

Ahmad, A., S. Senapati, M.I. Khan, R. Kumar and M. Sastry. 2003b. Extracellular biosynthesis of monodisperse gold nanoparticles by a novel extremophilic actinomycete, *Thermomonospora* sp. Langmuir 19(8): 3550–3553.

Ali, M.A., I. Rehman, A. Iqbal, S. Din, A.Q. Rao and A. Latif. 2014. Nanotechnology, a new frontier in agriculture. Adv. Life Sci. 1(3): 129–138.

Arya, V. 2010. Living systems: Eco-friendly nanofactories. Digest Journal of Nanomaterials and Biostructures 5(1): 9–21.

Ashajyothi, C., K. Jahanara and R.K. Chandrakanth. 2014. Biosynthesis and characterization of copper nanoparticles from *Enterococcus faecalis*. Int. J. Pharm. Bio. Sci. 5(4): 204–2011. doi:6(2)(B)36-52-April-2015.

Bai, H.J., Z.M. Zhang and J. Gong. 2006. Biological synthesis of semiconductor zinc sulphide nanoparticles by immobilized *Rhodobacter* sphaeroides. Biotechnol. Lett. 28(14): 1135–1139. doi:10.1007/s10529-006-9063-1.

Bai, H.J. and Z.M. Zhang. 2009. Microbial synthesis of semiconductor lead sulphide nanoparticles using immobilized *Rhodobacter* sphaeroides. Materials Letters 63(9-10): 764–766. doi:10.1016/j.matlet.2008.12.050.

Baker, S. and S. Shreedharmurthy. 2012. Antimicrobial activity and biosynthesis of nanoparticles by endophytic bacterium inhabiting *Coffee arabica* L. Scientific Journal of Biological Sciences 1(5): 107–113.

Balagurunathan, R., M. Radhakrishnan, R. Babu Rajendran and D. Velmurugan. 2011. Biosynthesis of gold nanoparticles by actinomycete *Streptomyces viridogens* strain HM10. Indian J. Biochem. Biophys. 48(5): 331–335.

Bhargava, A., N. Jain, M. Barathi, M.S. Akhtar, Y.S. Yun and J. Panwar. 2013. Synthesis, characterization and mechanistic insights of mycogenic iron oxide nanoparticles. J. Nanopart. Res. 15: 2031–2042.

Bhargava, A., N. Jain, M.A. Khan, V. Pareek, R.V. Dilip and J. Panwar. 2016. Utilizing metal tolerance potential of soil fungus for efficient synthesis of gold nanoparticles with superior catalytic activity for degradation of rhodamine B. J. Environ. Manag. 183: 22–32.

Bradfield, S.J., P. Kumar, J.C. White and S.D. Ebbs. 2017. Zinc, copper, or cerium accumulation from metal oxide nanoparticles or ions in sweet potato: Yield effects and projected dietary intake from consumption. Plant Physiol. Biochem. 110: 128–137. doi:10.1016/j.plaphy.2016.04.008.

Brayner, R., H. Barberousse, M. Hemadi, C. Djedjat, C. Yéprémian, T. Coradin, J. Livage, F. Fiévet and A. Coute. 2007. Cyanobacteria as bioreactors for the synthesis of Au, Ag, Pd, and Pt nanoparticles via an enzyme-mediated route. Journal of Nanoscience and Nanotechnology 7(8): 2696–2708. doi:10.1166/jnn.2007.600.

Cakmak, I. 2008. Enrichment of cereal grains with zinc: Agronomic or genetic biofortification? Plant and Soil 302(1-2): 1–17.

Chinnamuthu, C.R. and P.M. Boopathi. 2009. Nanotechnology and agroecosystem. Madras Agric. J. 96: 17–31.

Cunningham, D.P. and L.L. Lundie. 1993. Precipitation of cadmium by *Clostridium thermoaceticum*. Appl. Environ. Microbiol. 59(1): 7–14.

Das, S.K., A.R. Das and A.K. Guha. 2009. Gold nanoparticles: Microbial synthesis and application in water hygiene management. Langmuir 25(14): 8192–8199. doi:10.1021/la900585p.

Dhoke, S.K., P. Mahajan, R. Kamble and A. Khanna. 2013. Effect of nanoparticles suspension on the growth of mung (*Vigna radiata*) seedlings by foliar spray method. Nanotechnology Development 3(1): 1–5. doi:10.4081/nd.2013.e1.

Dhoondia, Z.H. and H. Chakraborty. 2012. *Lactobacillus* mediated synthesis of silver oxide nanoparticles. Nanomaterials and Nanotechnology. doi:10.5772/55741.

Dimetry, N.Z. and H.M. Hussein. 2016. Role of nanotechnology in agriculture with special reference to pest control. Int. J. Pharm. Tech. Res. 9: 121–144. doi:10.1007/s00253-012-3969-4.

Ditta, A. 2012. How helpful is nanotechnology in agriculture? Advances in Natural Sciences: Nanoscience and Nanotechnology 3(3): 033002. http://doi.org/10.1088/2043-6262/3/3/033002.

Du, L., H. Jiang, X. Liu and E. Wang. 2007. Biosynthesis of gold nanoparticles assisted by *Escherichia coli* DH5α and its application on direct electrochemistry of hemoglobin. Electrochemistry Communications 9(5): 1165–1170.

Duran, N., P.D. Marcato, M. Duran, A. Yadav, A. Gade and M. Rai. 2011. Mechanistic aspects in the biogenic synthesis of extracellular metal nanoparticles by peptides, bacteria, fungi, and plants. Appl. Microbiol. Biotechnol. 90: 1609–1624. doi:10.1007/s00253-011-3249-8.

Duran, N. and A.B. Seabra. 2012. Metallic oxide nanoparticles: State of the art in biogenic syntheses and their mechanisms. Appl. Microbiol. Biotechnol. 95(2): 275–288. http://doi.org/10.1007/s00253-012-4118-9.

Gade, A., A. Ingle, C. Whiteley and M. Rai. 2010. Mycogenic metal nanoparticles: Progress and applications. Biotechnol. Lett. 32: 593–600. doi:10.1007/s10529-009-0197-9.

Ghormade, V., M.V. Deshpande and K.M. Paknikar. 2011. Perspectives for nano-biotechnology enabled protection and nutrition of plants. Biotechnol. Adv. 29(6): 792–803. doi:10.1016/j.biotechadv.2011.06.007.

He, S., Z. Guo, Y. Zhang, S. Zhang, J. Wang and N. Gu. 2007. Biosynthesis of gold nanoparticles using the bacteria *Rhodopseudomonas capsulata*. Mater. Lett. 61(18): 3984–3987. doi:10.1016/j.matlet.2007.01.018.

Hennebel et al. 2011. Palladium nanoparticles produced by fermentatively cultivated bacteria as catalyst for diatrizoate removal with biogenic hydrogen. Appl. Microbiol. Biotechnol. 91(5): 1435–1445.

Husseiny, M.I. and M.A. El-Aziz, Y. Badr and M.A. Mahmoud. 2007. Biosynthesis of gold nanoparticles using *Pseudomonas aeruginosa*. Spectrochim. Acta Part A: Mol. Biomol. Spectrosc. 67(3-4): 1003–1006. doi:10.1016/j.saa.2006.09.028.

Iravani, S. 2014. Bacteria in nanoparticle synthesis: Current status and future prospects. Int. Sch. Res. Notices. 1–18. doi:10.1155/2014/359316.

Jain, N. and A. Bhargava, S. Majumdar, J.C. Tarafdar and J. Panwar. 2011. Extracellular biosynthesis and characterization of silver nanoparticles using *Aspergillus flavus* NJP08: A mechanism perspective. Nanoscale. 3: 635–641.

Jain, N. and A. Bhargava, J.C. Tarafdar, S.K. Singh and J. Panwar. 2013. A biomimetic approach towards synthesis of zinc oxide nanoparticles. Appl. Microbiol. Biotechnol. 97: 859–869. doi:10.1007/s00253-012-3934-2.

Javed, R. and M. Usman, B. Yucesan, M. Zia, and E. Gurel. 2017. Effect of zinc oxide (ZnO) nanoparticles on physiology and steviol glycosides production in micropropagated shoots of *Stevia rebaudiana* Bertoni. Plant Physiol. Biochem. 110: 94–99. doi:10.1016/j.plaphy.2016.05.032.

Jayaseelan, C. and A.A. Rahuman, A.V. Kirthi, S. Marimuthu, T. Santoshkumar, A. Bagavan, K. Gaurav, L. Karthik and K.V. Bhaskara Rao. 2012. Novel microbial route to synthesize ZnO nanoparticles using *Aeromonas hydrophila* and their activity against pathogenic bacteria and fungi. Spectrochim. Acta. A Mol. Biomol. Spectrosc. 90: 78–84.

Jeevan, P. and K. Ramya and A.E. Rena. 2012. Extracellular biosynthesis of silver nanoparticles by culture supernatant of *Pseudomonas aeruginosa*. Indian J. Biotechnol. 11: 72–76.

Jha, Z. and N. Behar, S.N. Sharma, G. Chandel, D.K. Sharma and M.P. Pandey. 2011. Nanotechnology: prospects of agricultural advancement. Nano Vision, 1(2): 88–100.

Kannan, N. and S. Subbalaxmi. 2010. Biogenesis of nanoparticles: A current perspective. Rev. Adv. Mater. Sci. 27: 99–114. http://doi.org/10.1254/jphs.09R03CP

Kathiresan, K., N.M. Alikunhi, S. Pathmanaban, A. Nabikhan and S. Kandasamy. 2010. Analysis of antimicrobial silver nanoparticles synthesized by coastal strains of *Escherichia coli* and *Aspergillus niger*. Can. J. Microbiol. 56(12): 1050–1059. Available at: http://www.nrcresearchpress.com/doi/abs/10.1139/W10-094.

Khot, L.R., S. Sankaran, J.M. Maja, R. Ehsani and E.W. Schuster. 2012. Applications of nanomaterials in agricultural production and crop protection: A review. Crop Prot. 35: 64–70. doi:10.1016/j.cropro.2012.01.007.

Klaus-Joerger, T., R. Joerger, E. Olsson and C.G. Granqvist. 2001. Bacteria as workers in the living factory: Metal-accumulating bacteria and their potential for materials science. Trends in Biotechnology 19: 15–20. doi:10.1016/S0167-7799(00)01514-6.

Klaus, T., R. Joerger, E. Olsson and C.G. Granqvist. 1999. Silver-based crystalline nanoparticles, microbially fabricated. Proc. Natl. Acad. Sci. 96: 13611–13614. doi:10.1073/pnas.96.24.13611.

Konishi, Y. et al. 2007. Bioreductive deposition of platinum nanoparticles on the bacterium *Shewanella algae*. J. Biotechnol. 128(3): 648–653.

Korbekandi, H., S. Iravani and S. Abbasi. 2012. Optimization of biological synthesis of silver nanoparticles using *Lactobacillus casei* subsp. *casei*. J. Chem. Technol. Biotechnol. 87: 932–937. doi:10.1002/jctb.3702.

Lengke, M.F., B. Ravel, M.E. Fleet, G. Wanger, R.A. Gordon and G. Southam. 2006. Mechanisms of gold bioaccumulation by filamentous cyanobacteria from gold (III)-chloride complex. Environmental Science and Technology 40(20): 6304–6309.

Lengke, M.F., M.E. Fleet and G. Southam. 2007. Biosynthesis of silver nanoparticles by filamentous cyanobacteria from a silver (I) nitrate complex. Langmuir 23(5): 2694–2699. doi:10.1021/la0613124.

Lengke, M.F., C. Sanpawanitchakit and G. Southam. 2011. Biosynthesis of gold nanoparticles: A review. Metal Nanoparticles in Microbiology. Springer, Berlin, Heidelberg. 37–74. doi: 10.1007/978-3-642-18312-6_1.

Li, X., H. Xu, Z.S. Chen and G. Chen. 2011. Biosynthesis of nanoparticles by microorganisms and their applications. J. Nanomater. doi:10.1155/2011/270974.

Mahajan, P., S.K. Dhoke and A.S. Khanna. 2011. Effect of nano-ZnO particle suspension on growth of mung (*Vigna radiata*) and gram (*Cicer arietinum*) seedlings using plant agar method. Journal of Nanotechnology. doi:10.1155/2011/696535.

Mandal, D., M.E. Bolander, D. Mukhopadhyay, G. Sarkar and P. Mukherjee. 2006. The use of microorganisms for the formation of metal nanoparticles and their application. Appl. Microbiol. Biotechnol. 69(5): 485–492. doi:10.1007/s00253-005-0179-3.

Marimuthu, S., A.A. Rahuman, A.V. Kirthi, T. Santhoshkumar, C. Jayaseelan and G. Rajakumar. 2013. Eco-friendly microbial route to synthesize cobalt nanoparticles using *Bacillus thuringiensis* against malaria and dengue vectors. Parasitology Research 112(12): 4105–4112. http://doi.org/10.1007/s00436-013-3601-2.

Matsunaga, T. and H. Takeyama. 1998. Biomagnetic nanoparticle formation and application. Supramolecular Science 5(3-4): 391–394. doi:10.1016/s0968-5677(98)00037-6.

Mi, C., Y. Wang, J. Zhang, H. Huang, L. Xu, S. Wang, X. Fang, J. Fang, C. Mao and S. Xu. 2011. Biosynthesis and characterization of CdS quantum dots in genetically engineered *Escherichia coli*. J. Biotechnol. 153(3-4): 125–132.

Minaeian, S., A.R. Shahverdi, A.A. Nouhi and H.R. Shahverdi. 2008. Extracellular biosynthesis of silver nanoparticles by some bacteria. Jundishapur J. Nat. Pharm. Prod. 17: 1–4.

Mishra, S. and H.B. Singh. 2014. Biosynthesized silver nanoparticles as a nanoweapon against phytopathogens: Exploring their scope and potential in agriculture. Appl. Microbiol. Biotechnol. 99: 1097–1107. doi:10.1007/s00253-014-6296-0.

Mishra, V.K. and A. Kumar. 2009. Impact of metal nanoparticles on the plant growth promoting Rhizobacteria. Digest Journal of Nanomaterials and Biostructures 4(3): 587–592.

Misra, A.N., M. Misra and R. Singh. 2013. Nanotechnology in agriculture and food industry. Int. J. Pure Appl. Sci. Technol. 16(2): 1–9.

Nangia, Y., N. Wangoo, N. Goyal, G. Shekhawat and C.R. Suri. 2009. A novel bacterial isolate *Stenotrophomonas maltophilia* as living factory for synthesis of gold nanoparticles. Microb. Cell Fact. 8: 1–7. doi:10.1186/1475-2859-8-39.

Popsecu, M., A. Velea and A. Lorinczi. 2010. Biogenic production of nanoparticles. Dig. J. Nanomater. Biostruct. 5(4): 1035–1040. doi:10.1002/ceat.200900046.

Posfai, M., B.M. Moskowitz, B. Arato, D. Schuler, C. Flies, D.A. Bazylinski and R.B. Frankel. 2006. Properties of intracellular magnetite crystals produced by *Desulfovibrio magneticus* strain RS-1. Earth Planet. Sci. Lett. 249(3-4): 444–455. doi:10.1016/j.epsl.2006.06.036.

Prasad, K. and A.K. Jha. 2009. ZnO nanoparticles: Synthesis and adsorption study. Natural Science 1(2): 129–135.

Prasad, R., V. Kumar and K.S. Prasad. 2014. Nanotechnology in sustainable agriculture: Present concerns and future aspects. Afr. J. Biotechnol. 13(6): 705–713. http://doi.org/10.5897/AJBX2013.13554.

Prasad, R., A. Bhattacharyya and Q.D. Nguyen. 2017. Nanotechnology in sustainable agriculture: Recent developments, challenges, and perspectives. Front. Microbiol. 8: 1–13. doi:10.3389/fmicb.2017.01014.

Priyadarshini, S. et al. 2013. Synthesis of anisotropic silver nanoparticles using novel strain, *Bacillus flexus* and its biomedical application. Colloids and Surf B: Biointerfaces. 102: 232–237. Available at: http://linkinghub.elsevier.com/retrieve/pii/S0927776512004663.

Pugazhenthiran, N., S. Anandan, G. Kathiravan, N.K.U. Prakash, S. Crawford and M. Ashokkumar. 2009. Microbial synthesis of silver nanoparticles by *Bacillus* sp. J. Nanopart. Res. 11(7): 1811–1815. http://doi.org/10.1007/s11051-009-9621-2.

Rai, M., A. Gade and A. Yadav. 2011. Biogenic nanoparticles: An introduction to what they are, how they are synthesized and their applications. *In*: Rai, M. and N. Duran (eds.). Metal Nanoparticles in Microbiology. Springer, Berlin, Heidelberg.

Rai, M. and A. Ingle. 2012. Role of nanotechnology in agriculture with special reference to management of insect pests. App. Microbiol. Biotechnol. 94(2): 287–293. doi:10.1007/s00253-012-3969-4.

Raliya, R. and J.C. Tarafdar. 2013. ZnO nanoparticle biosynthesis and its effect on phosphorous-mobilizing enzyme secretion and gum contents in Clusterbean (*Cyamopsis tetragonoloba* L.). Agric. Res. 2(1): 48–57. doi:10.1007/s40003-012-0049-z.

Raliya, R., P. Biswas and J.C. Tarafdar. 2015a. TiO$_2$ nanoparticle biosynthesis and its physiological effect on mung bean (*Vigna radiata* L.). Biotechnol. Rep. 5: 22–26. doi:10.1016/j.btre.2014.10.009.

Raliya, R., R. Nair, S. Chavalmane, W.N. Wang and P. Biswas. 2015b. Mechanistic evaluation of translocation and physiological impact of titanium dioxide and zinc oxide nanoparticles on the tomato (*Solanum lycopersicum* L.) Plant. Metallomics 7(12): 1584–1594. doi:10.1039/C5MT00168D.

Raliya, R., J.C. Tarafdar and P. Biswas. 2016. Enhancing the mobilization of native phosphorus in the mung bean rhizosphere using ZnO nanoparticles synthesized by soil fungi. J. Agric. Food Chem. 64(16): 3111–3118. doi:10.1021/acs.jafc.5b05224.

Ramalingam, V., R. Rajaram, C. Premkumar, P. Santhanam, P. Dhinesh, S. Vinothkumar and K. Kaleshkumar. 2014. Biosynthesis of silver nanoparticles from deep sea bacterium *Pseudomonas aeruginosa* JQ989348 for antimicrobial, antibiofilm and cytotoxic activity. J. Basic Microbiol. 54(9): 928–936. Available at: http://doi.wiley.com/10.1002/jobm.201300514.

Ramanathan, R., S.K. Bhargava and V. Bansal. 2011. Biological synthesis of copper/copper oxide nanoparticles. Chemeca 2011: Engineering a Better World: Australia, 18–21.

Ramanathan, R., R. Shukla, S.K. Bhargava and V. Bansal. 2015. Green synthesis of nanomaterials using biological routes. Nanomater. Environ. Prot. 329–348. doi:10.1002/9781118845530.ch20.

Rameshaiah, G.N., J. Pallavi and S. Shabnam. 2015. Nanofertilizers and nano sensors—An attempt for developing smart agriculture. Int. J. Eng. Res. Gen. Sci. 3(1): 314–320.

Ramezani, F., M. Ramezani and S. Talebi. 2010. Mechanistic aspects of biosynthesis of nanoparticles by several microbes. Nanocon. 10(12-14): 1–7.

Revati, K. and B.D. Pandey. 2011. Microbial synthesis of iron-based nanomaterials—A review. Bulletin of Materials Science 34(2): 191–198. Available at: http://link.springer.com/10.1007/s12034-011-0076-6.

Sabir, S., M. Arshad and S.K. Chaudhari. 2014. Zinc oxide nanoparticles for revolutionizing agriculture: Synthesis and applications. The Scientific World Journal. doi: 10.1155/2014/925494.

Saharan, V., R.V. Kumaraswamy, R.C. Choudhary, S. Kumari, A. Pal, R. Raliya and P. Biswas. 2016. Cu-chitosan nanoparticle mediated sustainable approach to enhance seedling growth in maize by mobilizing reserved food. J. Agric. Food Chem. 64(31): 6148–6155. doi:10.1021/acs.jafc.6b02239.

Saif Hasan, S., S. Singh, R.Y. Parikh, M.S. Dharne, M.S. Patole, B.L.V. Prasad and Y.S. Shouche. 2008. Bacterial synthesis of copper/copper oxide nanoparticles. J. Nanosci. Nanotechnol. 8(6): 3191–3196. doi:10.1166/jnn.2008.095.

Saifuddin, N., C.W. Wong and A.A. Yasumira. 2009. Rapid biosynthesis of silver nanoparticles using culture supernatant of bacteria with microwave irradiation. Journal of Chemistry 6(1): 61–70. doi:10.1155/2009/734264.

Sastry, M., A. Ahmad, M.I. Khan and R. Kumar. 2003. Biosynthesis of metal nanoparticles using fungi and actinomycete. Curr. Sci. 85(2): 162–170.

Sekhon, B.S. 2014. Nanotechnology in agri-food production: An overview. Nanotechnology, Science and Applications 7: 31–53.

Shahverdi, A.R., A. Fakhimi, H.R. Shahverdi and S. Minaian. 2007. Synthesis and effect of silver nanoparticles on the antibacterial activity of different antibiotics against *Staphylococcus aureus* and *Escherichia coli*. Nanomedicine: Nanotechnology, Biology and Medicine 3(2): 168–171. Available at: http://linkinghub.elsevier.com/retrieve/pii/S1549963407000469.

Shantkriti, S. and P. Rani. 2014. Biological synthesis of copper nanoparticles using *Pseudomonas fluorescens*. Int. J. Curr. Microbiol. App. Sci. 3(9): 374–383.

Sharma, N., A.K. Pinnaka, M. Raje, F.N.U. Ashish, M.S. Bhattacharyya and A.R. Choudhury. 2012. Exploitation of marine bacteria for production of gold nanoparticles. Microb. Cell Fact. 11(1): 86. doi:10.1186/1475-2859-11-86.

Sharon, M., A.K. Choudhary and R. Kumar. 2010. Nanotechnology in agricultural diseases and food safety. Journal of Phytology 2(4): 83–92.

Shi, X., C. Xue, F. Yu, T. Chen, H. Zhu, H. Xin and X. Wang. 2015. Functional nanomaterials engineered by microorganisms. Manufacturing Nanostructures 358–380.

Shobha, G., V. Moses and S. Ananda. 2014. Biological synthesis of copper nanoparticles and its impact—A review. Int. J. Pharm. Sci. Invent. 3(8): 2319–6718.

Singh, C.R., K. Kathiresan and S. Anandhan. 2015. A review on marine based nanoparticles and their potential applications. Afr. J. Biotechnol. 14(18): 1525–1532. doi: 10.5897/AJB2015.14527.

Singh, S., U.M. Bhatta, P.V. Satyam, A. Dhawan, M. Sastry and B.L.V. Prasad. 2008. Bacterial synthesis of silicon/silica nanocomposites. J. Mater. Chem. 18(22): 2601–2606.

Sintubin, L., W.E. Windt, J. Dick, J. Mast, D.V. Ha, W. Verstarete and N. Boon. 2009. Lactic acid bacteria as reducing and capping agent for the fast and efficient production of silver nanoparticles. Appl. Microbiol. Biotechnol. 84: 741–761.

Sivalingam, P., J.J. Antony, D. Siva, S. Achiraman and K. Anbarasu. 2012. Mangrove *Streptomyces* sp. BDUKAS10 as nanofactory for fabrication of bactericidal silver nanoparticles. Colloids and Surfaces B: Biointerfaces 98: 12–17. doi:10.1016/j.colsurfb.2012.03.032.

Srinivas, P.R., M. Philbert, T.Q. Vu et al. 2010. Nanotechnology research: Applications in nutritional aciences. The Journal of Nutrition. 140(1): 119–124. doi:10.3945/jn.109.115048.

Sunkar, S. and C.V. Nachiyar. 2012. Biogenesis of antibacterial silver nanoparticles using the endophytic bacterium *Bacillus cereus* isolated from *Garcinia xanthochymus*. Asian Pac. J. Trop. Biomed. 2(12): 953–959. doi:10.1016/S2221-1691(13)60006-4.

Sweeney, R.Y., C. Mao, X. Gao, J.L. Burt, A.M. Belcher, G. Georgiou and B.L. Iverson. 2004. Bacterial biosynthesis of cadmium sulfide nanocrystals. Chemistry & Biology 11(11): 1553–1559. http://doi.org/10.1016/j.

Tarafdar, J.C., R. Raliya, H. Mahawar and I. Rathore. 2014. Development of zinc nanofertilizer to enhance crop production in pearl millet (*Pennisetum americanum*). Agricultural Research 3(3): 257–262. http://doi.org/10.1007/s40003-014-0113-y.

Thakkar, K.N., S.S. Mhatre and R.Y. Parikh. 2010. Biological synthesis of metallic nanoparticles. Nanomedicine: Nanotechnology, Biology, and Medicine 6(2): 257–262. http://doi.org/10.1016/j. nano.2009.07.002.

Thamilselvi, V. and K.V. Radha. 2013. Synthesis of silver nanoparticles from *Pseudomonas putida* NCIM 2650 in silver nitrate supplemented growth medium and optimization using response surface methodology. Dig. J. Nanomater. Biostruct. 8(3): 1101–1111.

Vankova, R., P. Landa, R. Podlipna, P.I. Dobrev, S. Prerostova, L. Langhansova, A. Gaudinova, K. Motkova, V. Knirsch and T. Vanek. 2017. ZnO nanoparticle effects on hormonal pools in *Arabidopsis thaliana*. Sci. Total Environ. 59: 535–542. doi:10.1016/j.scitotenv.2017.03.160.

Vatta, L.L., R.D. Sanderson and K.R. Koch. 2006. Magnetic nanoparticles: Properties and potential applications. Pure Appl. Chem. 78(9): 1793–1801. doi:10.1351/pac200678091801.

Wei, X., M. Luo, W. Li, L. Yang, X. Liang, L. Xu, P. Kong and H. Liu. 2012. Synthesis of silver nanoparticles by solar irradiation of cell-free *Bacillus amyloliquefaciens* extracts and AgNO₃. Bioresour. Technol. 103(1): 273–278. http://doi.org/10.1016/j.biortech.2011.09.118.

Windt, W.D., P. Aelterman and W. Verstraete. 2005. Bioreductive deposition of palladium (0) nanoparticles on *Shewanella oneidensis* with catalytic activity towards reductive dechlorination of polychlorinated biphenyls. Environ. Microbiol. 7(3): 314–325.

Ya, S., S. Ab, U. Rb and L. Ramosa. 2015. An Overview of Nanotechnology Research Activities in the Agricultural Sector, 217–220.

CHAPTER 4

Perspectives of Nanomaterial in Sustainable Agriculture

S.K. Singh,* Rakesh Pathak and *Rajwant Kaur Kalia*

1. Introduction

The estimated world population is expected to grow to 9 billion by 2050 (UN 2015), which means that more food than what is currently being produced will be required. The adoption of efficient techniques will be helpful in making agriculture more sustainable (Chen and Yada 2011). The growing population and limited resources are the main constraints in making agriculture more viable and economic under the varied environmental conditions (Mukhopadhyay 2014). The agricultural development and growth is essential in reducing poverty and hunger and in promoting an increase in soil fertility (Campbell et al. 2014). Agricultural development also depends on the natural resources, climate change, social insertions, etc. All these factors should be addressed properly in order to achieve the oriented goal. More than 60% of the population of developing countries depend on agriculture for their livelihood (Brock et al. 2011), consequently, it is the most important sector for the economic health of these countries. The agriculture sector is facing various challenges, including food safety, production and risk of diseases, under the changing climatic conditions (Biswal et al. 2012); therefore, it is necessary to adopt a new technology that may provide better grain yield, productivity and sustainable growth of agriculture. In December 2002, the United States Department of Agriculture developed a draft for applying nanotechnology in the field of agriculture and food. It was suggested that agriculture should be automated, industrialized and reduced to simple functions. A number of nanomaterials have been developed with the help of nanotechnology and are being used in various fields including agriculture (Gibney 2015). Nanotechnology has the potential to transform the food industry by changing the way food is produced, processed,

ICAR- Central Arid Zone Research Institute, Jodhpur-342003, Rajasthan, India.
* Corresponding author

packaged, transported and consumed. Nanoscale science has the potential to provide novel and improved solutions to various problems being encountered in the field of agriculture (Chen and Yada 2011).

Nanotechnology is the science of very minute things but it also involves large things, comprising multi-disciplinary subjects, including Physics, Chemistry, Biology and other disciplines. Creation of materials/devices having new or almost different properties after manipulation in the atom, molecules or molecular clusters is the definition of nanotechnology in broad sense (Joseph and Morrison 2006); meanwhile, biotechnology uses the knowledge and techniques of biology for the development and services (Fakruddin et al. 2012).

Nanotechnological research is mainly focused in the field of electronics, energy, medicine and life sciences (Scrinis and Lyons 2007). Since agriculture is a comparatively non-potent industry, it is does not receive as much attention. Although some nano-chemical pesticides are in use, other products are at their early stages and they may take time to come into the reach of the common man. Application of nanotechnology has many challenges, including precise management of soil at micro level, targeted use of input, formulation of new toxins for the control of diseases, behaviour of crops under climate change scenarios, etc. (Prasad et al. 2014). The up-coming innovations in the agricultural field that are due to nanotechnology are quite interesting and promising. The quick detection of diseases and their management, as well as enhancements in the ability of plants to absorb nutrients or pesticides may be possible with nanotechnology (Tarafdar et al. 2013). The smart sensors and delivery systems developed with the help of nanotechnology will assist in combating viruses and other crop pathogens. These sensors have been designed in such a way that they are able to detect the presence of diseases prior to the appearance of symptoms. Simultaneously, the information will be relayed to the farmer and its remedy will be suggested. The work on nano-catalysts that will enhance the effectiveness of herbicides and pesticides leading to application of their lesser dosage is also taking place. The nanoparticles have been employed in various fields, including drug delivery (Lee et al. 2011), chemical and biological sensing (Barrak et al. 2016), gas sensing (Rawal and Kaur 2013, Mansha et al. 2016, Ullah et al. 2017), CO_2 capturing (Ganesh et al. 2013, Ramacharyulu et al. 2015) and other related applications (Shaalan et al. 2016). Various nanoparticles exhibit antibacterial, anticandidal, and antifungal activities (Aziz et al. 2016, Patra and Baek 2017).

At present, the sustainable production and efficiency of agriculture is inconceivable without the application of pesticides, fertilizers, etc. However, these agrochemicals have various associated issues, including contamination of water or their residues on food products threatening human and environmental health. The precise management and control of inputs may reduce these risks (Kah 2015). The high-tech agricultural system and engineered smart nano-tools could be the best strategy to make a revolution in the agricultural practices and to enhance the quality and quantity of yields (Liu and Lal 2015). The present review emphasises the perspective of nanomaterials in sustainable agriculture.

2. Challenges and Limitation of Conventional Farming

The green revolution has significantly increased the grain yield in order to fulfil the global food requirement. Nevertheless, its disadvantageous impacts on the environment and ecosystem are now clearly visible, suggesting the requirement for adoption of comparatively more sustainable agricultural approaches (Tillman et al. 2002). The extreme and unbalanced application of fertilizers and pesticides has disturbed the nutrient levels and increased toxins in groundwater and surface waters. It not only degraded the soil quality but also increased the expense of fertilization, irrigation and energy to maintain productivity of the land (IFPRI 2002). The groundwater levels are decreasing fast in areas where more water is being pumped out for irrigation (Rodell et al. 2009), and around 40% of crop production comes only from the 16% of irrigated agricultural fields (Mukhopadhyay 2014). Besides this, the long-term irrigation and drainage practices have increased the speed of weathering of soil minerals, making soils acidic and damaging the carbon profile in soils (Knorr et al. 2005). All these situations have resulted in the abandonment of some of the best farming lands (Mukhopadhyay 2005).

There are number of challenges associated with conventional farming, hence, various approaches of alternative farming, like conservation agriculture and organic farming, have come into existence. However, these approaches are not able to achieve higher production and productivity (Kirchmann and Thorvaldsson 2000). Therefore, it is high time to re-engineer plants (Eapen and D'Souza 2005) and nanomaterials may be a useful tool in coping with the problems.

3. Advantages of Nanomaterials Over Corresponding Bulk Materials

The bulk materials show unusual characteristics at nanoscale in terms of surface area, cation exchange capacity, ion adsorption and complexation. Nanomaterials have a high proportion of atoms (Maurice and Hochella 2008), different surface compositions, density sites and reactivities with respect to adsorption and redox reactions (Waychunas et al. 2005, Hochella et al. 2008). These advantages of the nanomaterials can be exploited for their application in agriculture.

4. Properties of the Nanoparticles

The increased relative surface area and quantum effects are the main factors that make a difference between the nanomaterials and other materials. Nanoparticles are not simple molecules and are composed of three layers, i.e., surface layer, shell layer and core. The surface layer is functionalized with small molecules, metal ions, surfactants and polymers. The shell layer is a chemically different material from the core, while the core is the central portion and is referred to as the nanoparticle itself (Shin et al. 2016). The important properties of the nanoparticles include hydrophobicity, solubility-release of toxic species, surface area/roughness, surface species contaminations/adsorption, reactive oxygen species (ROS), capacity to produce ROS, structure/composition, competitive binding sites with receptor and dispersion/aggregation, etc. (Somasundaran et al. 2010).

5. Agricultural Development Through Nanomaterials

The environment will be indirectly protected through the use of alternative or renewable energy supplies and filters or catalysts to reduce pollution with the use of nanotechnology. The controlled environment agriculture (CEA) is an agricultural system that utilizes modern technologies for crop management and provides a platform to initiate the use of nanotechnology in agriculture. The automated system monitors and controls localised environments for crops. The smart sensors and delivery systems will combat viruses and other crop pathogens in the agricultural field.

Economically effective nanomaterials play an important role in the purification of irrigation water. The membrane filters developed from carbon nanotubes, nanoporous ceramics and magnetic nanoparticles are common in water treatment (Hillie and Hlophe 2007). The carbon nanotubes are able to remove water-borne pathogens, heavy metals and toxicants from the potable water, while positive charged nanoceramic filter are able to trap the bacteria and viruses (Argonide 2005).

Arsenic can be removed with the help of simple synthetic clay nanomineral by percolating through a hydrotalcite column. Besides this, Zinc oxide nanoparticles can also be used to remove the arsenic impurity from the water. Other nanomaterials that could be used in the remediation of harmful pollutants include nanoscale zero valent iron, zeolites, oxides of metals, carbon nanofibers, enzymes and titanium dioxide. Reports suggest that a variety of nanoparticle filters have been used to remove organic particles and pesticides, viz., DDT, Endosulfan, Malathion and Chlorpyrifos from water (Karn et al. 2009). Slow release nano-encapsulated fertilizers have been widely used to minimize fertilizer consumption and environmental pollution (DeRosa et al. 2010).

6. Application of Nanomaterial in Agriculture

The material at nano-level behaves differently and shows different properties. The nano-forms of carbon, silver, silica and alumina-silicates have been found to control plant diseases. The silver nanoparticles have been used as antimicrobial agents and prevent the expression of ATP production (Yamanka et al. 2005). A brief account of the application of different nanomaterials in agriculture is summarized below.

6.1 Nano-Encapsulation

The encapsulation technique is getting more attention in the field of agriculture due to its unique quality of encapsulation of pesticides, fertilizer or other ingredient inside the nanoparticles that can be timely released or have release linked to an environmental trigger. In this technique, the outer shell of a capsule is manipulated in such a way as to control the release of the substance to be delivered. The controlled release strategy has got tremendous acceptance in the medical field. Nano-encapsulation of nanoparticles in the form of pesticides allows the appropriate absorption of the chemicals into the plants (Scrinis and Lyons 2007). Through this process, desired chemicals can also be transported into the plant tissues for the protection of plants against insects and diseases (Torney 2009). It is one of the most potentially effective techniques to protect the plant against insects and pests because it works through diffusion, dissolution, biodegradation and osmotic pressure (Ding and Shah 2009, Vidhyalakshmi et al. 2009). Therefore, formulation of nanoscale pesticides through nano-encapsulation is receiving more attention. The carbon nanotubes can be used as regulators of seed germination and plant growth (Khodakovskaya et al. 2012).

6.2 Nano-Fertilizers

As per estimates, around 40–70% of nitrogen, 80–90% of phosphorus, and 50–90% of potassium are lost in the environment due to leaching of chemicals, drift, runoff, evaporation, hydrolysis by soil moisture, as well as photolytic and microbial degradation, and could not reach the plant (Trenkel 1997, Ombodi and Saigusa 2000). To overcome the lost quantities, repeated application of fertilizer and pesticide that adversely affects the nutrient balance of the soil as well as the economy is taking place. The excess application of fertilizers and pesticides also leads to increased pathogens, pest resistance, bioaccumulation of pesticides, reduced soil microflora, diminished nitrogen fixation, and destroys bird habitats (Tilman et al. 2002). Nanotechnology has provided the possibility of developing nanoparticles as fertilizer carriers or controlled-release vectors to improve the nutrient use efficiency and reduce the environmental pollution (Chinnamuthu and Boopati 2009). The development and application of nanofertilizers has smart delivery systems for the transportation of nutrients due to advanced properties like size, surface area, etc.

In agriculture, synthetic agrochemicals need to be replaced by biopolymer materials in order to reduce the deleterious effects of resistance in plant pathogens, harm to non-target organisms and deterioration of soil health (Kashyap et al. 2015, Zhan et al. 2015). An NPK composition fertilizer, having an inner coating of chitosan and outer coating of acrylic acid-co acrylamide (P(AA-co-AM)), was developed (Wu et al. 2008). The product had slow controlled release of nutrients, and neither the matrix polymers nor their degraded products were destructive to the soil. Besides the slow release of the fertilizers, chitosan nanoparticles also have antibacterial properties (Corradini et al. 2010). Chitosan was found to be the best matrix to carry the essential oil of *Lippia sidoides*, which has insecticidal properties (Paula et al. 2011), and it controlled the proliferation of insect larvae (Kashyap et al. 2015). The microcapsules of alginate and chitosan were also evaluated as a carrier system for imidacloprid and it was observed that the release time of the encapsulated insecticide was up to eight times higher, as compared to the free insecticide (Guan et al. 2008). Chitosan, a biocompatible and non-toxic biodegradable material, is being exploited in agriculture (Katiyar et al. 2015). The application of Cu-chitosan NPs significantly controlled Curvularia leaf spot disease and had a growth promoting effect in maize (Saharan et al. 2016, Choudhary et al. 2017).

Nano-fertilizers may contain nano zinc, silica, iron and titanium dioxide. The ZnCdSe/ZnS core shell quantum dots (QDs), InP/ZnS core shell QDs, Mn/ZnSe QDs, gold nanorods, core shell QDs, etc., may also be associated with the Nano-fertilizers. The toxicity, absorbance and biological fate of various nanoparticles, viz. Al_2O_3, TiO_2, CeO_2, FeO and ZnO, have been studied for agricultural production (Dimkpa 2014, Zhang et al. 2016). Various workers have also studied the size, degree of aggregation and zeta potential of metal oxide nanoparticles in the presence of proteins and cell media (Llop et al. 2014, Marzbani et al. 2015).

TiO_2 has a significant effect on the growth of corn and also enhances light absorption and photo-energy transfer (Lu et al. 2002). Enhanced nitrate reductase activity and plant absorption capacity in soybean have also been reported with SiO_2 and TiO_2 nanoparticles (Lu et al. 2002). It was observed that that SiO_2 nanoparticles improved germination in tomato seeds (Manzer and Mohamed 2014, Ramalingam et al. 2015). Biosynthesized MgO nanoparticle enhanced chlorophyll content in the leaves of *Cyamopsis tetragonoloba* (Raliya et al. 2014a). ZnO nanoparticle induced exopolysaccharide (EPS) production from Bacillus subtilis and also improved soil aggregation, moisture retention and soil organic carbon (Raliya et al. 2014b). The nanoporous zeolites also had increased efficiency and slow release of fertilizers (Chinnamuthu and Boopathi 2009).

The yield and growth of crops can be improved with the help of zinc oxide nanoparticles. Studies revealed that the peanut seeds treated with different concentrations of zinc oxide nanoparticles showed increased seed germination, seedling vigor and plant growth, and the NPs proved to be effective in increasing stem and root growth (Prasad et al. 2012a). Similarly, seeds of wheat treated with metal nanoparticles gave 20–25% higher yields (Batsmanova et al. 2013). It was observed that the nano-calcium carbonate leads to an increase in the number of leaves and leaf area, dry weight, the soluble sugar and peanut protein content (Liu et al. 2009). The nanopowders can be successfully used as fertilizers and pesticides (Selivanov and Zorin 2001, Rusonik et al. 2003).

It has been observed that, with the application of nanofertilizers, higher yield can be obtained and the nutrients retain their activity for a longer period of time (Naderi and Danesh-Shahraki 2013, Ditta et al. 2015, Ditta and Arshad 2015) as compared to traditional fertilizers. Nano-fertilizers might be the best substitute for ordinary fertilizers as they release the nutrients as per requirement of the crop and provide stress tolerating ability, as well as increasing the nutrient use efficiency three-fold. The surface coating of nanomaterials on fertilizer particles hold the material more strongly due to higher surface tension in comparison to conventional surfaces. The toxic effects associated with the overdosing of the fertilizers may be controlled and the soil quality can be improved with the application of nano-fertilizers (Suman et al. 2010). Since nutrients and growth promoters are encapsulated at nanoscale in the nanofertilizers, they are used efficiently at slow and targeted pace. The nanofertilizers are not crop specific and help to maintain the carbon uptake of the soil and improve soil aggregation in an eco-friendly manner. As compared to chemical fertilizers, nanofertilizers are required in lesser amount and are cheaper.

6.3 Nano-Herbicide

Application of mild susceptible herbicide in one season and another herbicide in another season develops resistance, shift in weed flora and sometimes becomes uncontrollable through chemicals due to its continuous use. Similarly application of single herbicide for multispecies crops also results in poor control and herbicide resistance. The excess herbicides leave residue in the soil and may spoil the succeeding crops. Therefore, development of target specific herbicide is necessary in order to reduce the dosage of herbicide and enhance its efficacy. The nano-herbicides will remove the weeds without leaving any toxic residues and will prevent the growth of weed species. If the active ingredients of the herbicides are combined with the smart delivery system, they can be applied according to the requirement and the condition of the field. It will improve not only the efficacy of the herbicide but also reduce the soil contamination and harmful effect to the agriculture workers. The herbicide molecule encapsulated with nanoparticle is target specific. The root-specific receptor herbicides enter into the root system and starve the specific weed plant of food until it dies (Chinnamuthu and Kokiladevi 2007). The weeds also have great competition with the crop for nutrient, moisture, etc., leading to yield loss, particularly in rain-fed areas.

Herbicides cannot be used in advance before the germination of crop. In this situation, controlled release of encapsulated herbicides may take care of the competing weeds with crops. Susha et al. (2009) applied silver modified with nanoparticles of magnetite stabilized with Carboxy Methyl Cellulose (CMC) nanoparticles and recorded 88% degradation of herbicide atrazine residue under controlled environment. Silva et al. (2011) prepared alginate/chitosan nanoparticles as a carrier system for the herbicide paraquat. They reported significant differences between the release profiles of free paraquat and the herbicide one associated with the alginate/chitosan nanoparticles. The association of paraquat with alginate/chitosan nanoparticles also altered the release profile of the herbicide and its interaction with the soil.

6.4 Nanobiosensors

Nanosensors have the ability to monitor the soils and water contamination (Ion et al. 2010); similarly, biosensors, electrochemical sensors, optical sensors will be the choice of instruments for detecting the heavy metals in trace range (Ion et al. 2010). Sensors give better results, live pictures and conditions of the field (Ibtisam 2001) and monitor changes or the effects caused by various pesticides, fertilizers and herbicide. They are able to record the soil pH, moisture level and growth conditions of crop, stem fruit or even root. They can also monitor the toxicity produced in the field. The sensors start detecting all these parameters and alert the farmer to take appropriate measures accordingly.

The sensitivity and performance of biosensors can be enhanced with the help of nanomaterials using signal transduction technologies (Sagadevan and Periasamy 2014). Presently, nanotechnology-based biosensors are in the budding phase (Fogel and Limson 2016) but it is having immense applications in various fields including agriculture. The advancement of various procedures and tools required for the design and development of nanomaterials is rapidly increasing regarding the assessment of various metals (gold, silver, cobalt, etc.), to be used as potentials biosensors. The study of biosensors is opening an avenue between biological and material science. It combines a biological recognition element based on the physical or chemical principles. It is an integration of biological and electronic components where various biological signals are processed and measured through an electronic device. Brolo (2012) developed a micro cantilever-based DNA biosensor with the help of gold nanoparticles that is being widely used to detect low level DNA concentrations during a hybridization reaction. The use of nano-sensors disseminated in the field with water, nutrient and chemicals during the farming enables the farmers to detect the existence of pests, levels of soil nutrients and stress conditions in the crop (Ingale and Chaudhari 2013). The study of regulation of plant growth hormones resulted in to the development of nano-sensors reacting with the auxin. This is a step forward in understanding how the roots of plant adjust to their environment (McLamore et al. 2010).

The sensors received the responses from the physico-chemical and biological aspects and transmit them into a signal (NNCO 2009). They are able to detect the microbial contaminants, nutrient content and plant stress appearing the crop and also indicate the nutrient and water status of the crop plant. The information will be of great significance for the better decision of the farmer and they will be able to identify the area of field requiring the nutrient, water, insecticide or pesticide, etc.

6.5 Nano Pesticide

Pesticides are used to bring down the pest population below the economic threshold level at various stages of crop growth for a longer period. So, the application of active ingredients is the most important issue in terms of cost and methods under varied environmental conditions. The application of nanomaterials to plant protection is an under-explored area. The application of nano-biopesticides is more suitable as they are safe for plants and are eco-friendly as compared to conventional biopesticides (Barik et al. 2008). The nanoparticle of various metals have been found as an effective alternative against various pathogens (Prasad et al. 2011, Swamy and Prasad 2012, Prasad and Swamy 2013) and have great potential in the management of plant diseases (Park et al. 2006). The nano-encapsulation technique provides the best and most precise solution for this problem. The nanoencapsulated pesticide formulation has slow releasing properties with improved solubility, specificity, permeability and stability (Bhattacharyya et al. 2016). The nano-sized particles of active ingredients of insecticides, fungicides or nematicides sealed in a thin walled protective coating will produce a formulation that will offer effective control of pests. The nanoencapsulated pesticides require reduced dosage of pesticides, leading to reduced cost and a lower environmental impact on the ecosystem (Nuruzzaman et al. 2016). These pesticides may be time released or released upon the occurrence of an environmental trigger. Simultaneously, its residue will not accumulate in the soil. The application of nano-pesticides will reduce the frequency of its application and quantity by 10–15 times as compared to classical formulations. Microencapsulated pesticides are being marketed by various chemical companies with different brand names (Gouin 2004). Clay nanotubes have been developed as carriers of pesticides for extended release and better contact with plants. These tubes reduce the amount of pesticides by 70–80%, thereby reducing the cost of pesticide with minimum impact on water streams.

Poultry feed containing nanoparticles help in decreasing food-borne pathogens. Nanomaterials can be applied as photocatalysts in agriculture. Various nanostructures of titanium dioxide and zinc oxide have been widely studied as photocatalysts (Ullah and Datta 2008). Nano-titanium dioxide having self-sanitizing photocatalyst coating used in poultry could be applied to oxidize and destroy the bacteria. The photocatalytic properties of nano-titanium dioxide are activated in the presence of natural or UV lights. The surface remains self-sanitizing after the coating and remains active until the presence of enough light such that the photocatalytic effect is activated. The Chicken and Hen Infection Program in Denmark included the self-cleaning and disinfection nanocoatings (Clemants 2009). With the view that the nanoscale smooth surface acts as more effective disinfection and cleaning, research in the area is getting attention and researchers are working on a coating with nanosilver that does not require UV light for activation. Studies revealed that the nanoparticles could stimulate fibroin protein production in silk worms (Bhattacharyya 2009).

Nanoparticles are used in the preparation of new formulations, like pesticides, insecticides and insect repellants (Barik et al. 2008, Gajbhiye et al. 2009), and are effective against insects and pests. The nano-emulsions (Wang et al. 2007) and essential oil-loaded solid lipid nanoparticles used in the formulation of pesticides could be effective against various insects and pests (Liu et al. 2006).

Nanosilica can be successfully used as a nanopesticide (Barik et al. 2008). The insect pests use cuticular lipids to protect their water barrier and prevent death from dehydration. The nanosilica spread on the plant surface come into contact with the insect and are absorbed into the cuticular lipids causing the death of the insect. Ulriches et al. (2005) reported that the modified surface-charged hydrophobic nanosilica may be applied to manage a variety of ectoparasites of animals and agricultural insect pests. The insecticidal activity of poly-ethylene glycol-coated nanoparticles loaded with garlic essential oil has been observed against adult *Tribolium castaneum* insect and the control efficacy was found to be 80%, revealing the slow release of the active components from the nanoparticles (Yang et al. 2009).

Various nanoparticles, i.e., silver nanoparticles, aluminium oxide, zinc oxide and titanium dioxide, have been found to be effective in the management of rice weevil and grasserie disease in silk worm (Goswami et al. 2010). Significant insecticidal activity of nanostructured alumina was observed against the major insect pests associated with stored food, viz., *Sitophilus oryzae* and *Rhyzopertha dominica* (Teodoro et al. 2010). The nanoalumina may be a reliable alternative for the control of insect pests in the disease management of stored food.

The nanoparticles of zinc oxide and magnesium oxide have antibacterial efficacy (Shah and Towkeer 2010) and have also been recommended as an antimicrobial preservative for wood and food products (Aruoja et al. 2009, Sharma et al. 2009). Their optical transparency, greater dispensability and smoothness makes them an ideal choice of antibacterial ingredients in many products. Suman et al. (2010) reported that the silver nanoparticles have high surface area and high fraction of surface atoms and antimicrobial effects compared to the bulk. Nanosilver has been the most studied nanoparticle and is known for strong inhibitory and bactericidal effects, along with broad spectrum antimicrobial activities (Swamy and Prasad 2012, Prasad et al. 2012b, Prasad and Swamy 2013). Besides this, the fungicidal properties of nano-size silver colloidal solution has also been reported against various plant pathogens (Kim et al. 2012).

7. Ecotoxicological Aspects of Nanoparticles

With the progression of the technology, synthesis of nanoparticles is increasing significantly, leading us to consider their antagonistic effect on human and plant health. Various studies have been undertaken in order to assess the toxicity of the nanoparticles generally used in various fields (Rana and Kalaichelvan 2013, Du et al. 2017, Tripathi et al. 2017a,b,c). The nanotoxicity may be attributed to the electrostatic interaction of nanoparticles with the membrane and to their deposition in the cytoplasm (Aziz et al. 2015, 2016). Some of the nanoparticles, viz., TiO_2, ZnO, SiO_2, etc., are photochemically active and excite electrons upon exposure to light and form superoxide radicals in the presence of oxygen (Hoffmann et al. 2007). The number of protective enzymatic or genetic constitutions of the cell increases as the result of oxidative stress and can be estimated in the context of toxicity and ecotoxicity (Vannini et al. 2014). It is essential to determine the hindered influences of environmental exposure to nanoparticles in order to assess the potential adaptive approaches (Cox et al. 2017, Singh et al. 2017). The nanoparticles enter through cellular system in the plant, translocate in shoot and accumulate in different aerial parts of the plants. The accumulated nanoparticles amend the process of photosynthesis and affect the translocation of food material (Shweta et al. 2016, Tripathi et al. 2016a, Du et al. 2017). The accumulation of the nanoparticles of both the ferrous and ferric oxides has been observed naturally in *Lepidium sativum* and *Pisum sativum* plants, showing the presence of nanoparticles in the natural ecosystem (Abbas et al. 2016). The ecotoxicity establishes a direct link between nanoparticles and the organism at various trophic levels (Tripathi et al. 2016b), therefore, its assessment is essential. The inclusion of artificially synthesized (either chemical or green) nanoparticles in the agricultural field should be checked properly with regard to bio-safety.

Conclusions and Future Perspectives

Agriculture is the major industry that provides food for human beings. Modern knowledge in the field of agriculture is necessary in order to increase production and efficiency. Nanotechnology allows for better management and conservation of inputs in agriculture (Sugunan and Dutta 2008). The extensive use of agrochemicals to increase agricultural production has polluted the top soil, groundwater and food. Increase in the agricultural productivity is essential but the health of soil and the environment is also necessary. Therefore, exploration of new approaches is vital to increase the agricultural productivity coupled with environmental health. The application of nanoparticles for the supply of pesticides and fertilizers is expected to reduce the dosage of these agrochemicals and ensure controlled slow delivery. Since nanomaterials have increased surface area, they might have toxic effects (Nel et al. 2006), therefore, selection of nano-materials for application in the field is crucial. The toxicity level of the nanoparticles should be defined properly, along with the risk assessment and its adverse effect on human health. Although integration of any new technology in the system is always apprehended with the unanticipated threats, awareness about advantages and challenges associated with the technology is essential. The development of regulation, legislation, databases and alarm systems is required in order to exploit the benefits of this budding technology. In the long term, this technology may result in economic development worldwide for quality and quantity of agricultural produce. The impact of nanomaterials on human beings, the economy and agricultural science should be debated with all the stakeholders, including researchers, authorities, industrial sectors and farmers.

References

Abbas, S.S., M. Haneef, M. Lohani, H. Tabassum and A.F. Khan. 2016. Nanomaterials used as a plants growth enhancer: An update. Int. J. Pharm. Sci. Rev. Res. 5: 17–23.

Argonide. 2005. NanoCeram filters. Argonide Corporation. http://sbir.nasa.gov/SBIR/ successes/ ss/9-072text.html.

Aruoja, V., H. Dubourguier, C. Kasamets and K.A. Kahru. 2009. Toxicity of nanoparticles of CuO, ZnO and TiO$_2$ to microalgae, *Pseudokirchneriella subcapitata*. Sci. Total Environ. 407: 1461–1468.

Aziz, N., M. Faraz, R. Pandey, M. Sakir, T. Fatma, A. Varma, I. Barman and R. Prasad. 2015. Facile algae-derived route to biogenic silver nanoparticles: Synthesis, antibacterial and photocatalytic properties. Langmuir 31: 11605–11612.

Aziz, N., R. Pandey, I. Barman and R. Prasad. 2016. Leveraging the attributes of Mucor hiemalis-derived silver nanoparticles for a synergistic broad-spectrum antimicrobial platform. Front. Microbiol. 7: 1984. doi: 10.3389/fmicb.2016.01984.

Barik, T.K., B. Sahu and V. Swain. 2008. Nanosilica-from medicine to pest control. Parasitol. Res. 103: 253–258.

Barrak, H., T. Saied, P. Chevallier, G. Laroche, A. M'nif and A.H. Hamzaoui. 2016. Synthesis, characterization, and functionalization of ZnO nanoparticles by N-(trimethoxysilylpropyl) ethylenediamine triacetic acid (TMSEDTA): Investigation of the interactions between phloroglucinol and ZnO@TMSEDTA. Arab. J. Chem. http://dx.doi.org/10.1016/j.arabjc. 2016.04.019.

Batsmanova, L.M., L.M. Gonchar, N.Y. Taran and A.A. Okanenko. 2013. Using a colloidal solution of metal nanoparticles as micronutrient fertiliser for cereals. Proceedings of the International Conference Nanomaterials, Alushta, the Crimea, A. Pogrebnjak (Ed.), Sumy State University, 2013. - V.2, No4. - 04NABM14.

Bhattacharyya, A. 2009. Nanoparticles from drug delivery to insect pest control. Akshar 1(1): 1–7.

Bhattacharyya, A., P. Duraisamy, M. Govindarajan, A.A. Buhroo and R. Prasad. 2016. Nano-biofungicides: Emerging trend in insect pest control. pp. 307–319. *In*: Prasad, R. (ed.). Advances and Applications through Fungal Nanobiotechnology. Cham: Springer International Publishing. doi:10.1007/978-3-319-42990-8_15.

Biswal, S.K., A.K. Nayak, U.K. Parida and P.L. Nayak. 2012. Applications of nanotechnology in agriculture and food sciences. Int. J. Sci. Innovat. Discov. 2(1): 21–36.

Brock, D.A., T.E. Douglas, D.C. Queller and J.E. Strassmann. 2011. Primitive agriculture in a social amoeba. Nature 469: 393–396.

Brolo, A.G. 2012. Plasmonics for future biosensors. Nat. Photonics 6: 709–713.

Campbell, B.M., P. Thornton, R. Zougmoré, P. van Asten and L. Lipper. 2014. Sustainable intensification: What is its role in climate smart agriculture? Curr. Opin. Environ. Sustain. 8: 39–43.

Chen, H. and R. Yada. 2011. Nanotechnologies in agriculture: New tools for sustainable development. Trends Food Sci. Technol. 22(11): 585–594.

Chinnamuthu, C.R. and E. Kokiladevi. 2007. Weed management through nanoherbicides. *In*: C.R. Chinnamuthu, B. Chandrasekaran and C. Ramasamy (eds.). Application of Nanotechnology in Agriculture. Tamil Nadu Agricultural University, Coimbatore, India.

Chinnamuthu, C.R. and P.M. Boopathi. 2009. Nanotechnology and Agroecosystem. Madras Agric. J. 96: 17–31.

Choudhary, R.C., R.V. Kumaraswamy, S. Kumari, S.S. Sharma, A. Pal, R. Raliya, P. Biswas and V. Saharan. 2017. Cu-chitosan nanoparticle boost defense responses and plant growth in maize (*Zea mays* L.). Scientific Reports 7: 9754. doi:10.1038/s41598-017-08571-0.

Clemants, M. 2009. Pullet production gets silver lining. Poultry International.

Corradini, E., M.R. Moura and L.H.C. Mattoso. 2010. A preliminary study of the incorporation of NPK fertilizer into chitosan nanoparticles express. Polymer Lett. 4: 509–15.

Cox, A., P. Venkatachalam, S. Sahi and N. Sharma. 2017. Reprint of: Silver and titanium dioxide nanoparticle toxicity in plants: A review of current research. Plant Physiol. Biochem. 110: 33–49.

DeRosa, M.C., C. Monreal, M. Schnitzer, R. Walsh and Y. Sultan. 2010. Nanotechnology in fertilizers. Nat. Nanotechnol. 5: 91–94.

Dimkpa, C.O. 2014. Can nanotechnology deliver the promised benefits without negatively impacting soil microbial life? J. Basic Microbiol. 54: 889–904.

Ding, W.K. and N.P. Shah. 2009. Effect of various encapsulating materials on the stability of probiotic bacteria. J. Food Sci. 74(2): M100–M107.

Ditta, A. and M. Arshad. 2015. Applications and perspectives of using nanomaterials for sustainable plant nutrition. Nanotechnol. Rev. 5(2): 209–230.

Ditta, A., M. Arshad and M. Ibrahim. 2015. Nanoparticles in sustainable agricultural crop production: Applications and Perspectives. pp. 55–75. *In:* M.H. Siddiqui, M.H. Al-Whaibi, F. Mohammad (eds.). Nanotechnology and Plant Sciences-Nanoparticles and their impact on Plants. Springer, Switzerland.

Du, W., W. Tan, J.R. Peralta-Videa, J.L. Gardea-Torresdey, R. Ji, Y. Yin and H. Guo. 2017. Interaction of metal oxide nanoparticles with higher terrestrial plants: Physiological and biochemical aspects. Plant Physiol. Biochem. 110: 210–225.

Eapen, S. and S.F. D'Souza. 2005. Prospects of genetic engineering of plants for phytoremediation of toxic metals. Biotechnol Adv. 23: 97–114.

Fakruddin, Md., Z. Hossain and H. Afroz. 2012. Prospects and applications of nanobiotechnology: A medical perspective. J. Nanobiotechnol. 10: 31.

Fogel, R. and J. Limson. 2016. Developing biosensors in developing countries: South Africa as a case study. Biosensors 6: 5. doi:10.3390/bios6010005.

Gajbhiye, M., J. Kesharwani, A. Ingle, A. Gade and M. Rai. 2009. Fungus mediated synthesis of silver nanoparticles and its activity against pathogenic fungi in combination of fluconazole. Nanomedicine 5(4): 282–286.

Ganesh, M., P. Hemalatha, M.M. Peng and H.T. Jang. 2013. One pot synthesized Li, Zr doped porous silica nanoparticle for low temperature CO_2 adsorption. Arab. J. Chem. http://dx.doi.org/10.1016/j.arabjc.2013.04.031.

Gibney, E. 2015. Buckyballs in space solve 100-year-old riddle. Nature News doi: 10.1038/nature.2015.17987.

Goswami, A., I. Roy, S. Sengupta and N. Debnath 2010. Novel applications of solid and liquid formulations of nanoparticles against insect pests and pathogens. Thin Solid Films 519: 1252–1257.

Gouin, S. 2004. Microencapsulation: Industrial appraisal of existing technologies and trends. Trends Food Sci. Technol. 15: 330–347.

Guan, H., D. Chi, J. Yu and X. Li. 2008. A novel photodegradable insecticide: Preparation, characterization and properties evaluation of nano-imidacloprid. Pestic. Biochem. Physiol. 92: 83–91.

Hillie, T. and M. Hlophe. 2007. Nanotechnology and the challenge of clean water. Nat. Nanotechnol. 2: 663–664.

Hochella, M.F. Jr, S.K. Lower, P.A. Maurice, R.L. Penn, N. Sahai, D.L. Sparks and B.S. Twining. 2008. Nanominerals, mineral nanoparticles, and earth systems. Sci. 319: 1631–1635.

Hoffmann, M., E.M. Holtze and M.R. Wiesner. 2007. Reactive oxygen species generation on nanoparticulate material. pp. 155–203. *In*: Wiesner, M.R. and J.Y. Bottero (eds.). Environmental Nanotechnology. Applications and Impacts of Nanomaterials, New York McGraw Hill.

Ibtisam, E.T. 2001. Biosensors developments and potential applications in the agricultural diagnosis sector. Elsevier Sci. Computers Electronics Agric. 30(1): 205–218.

[IFPRI] International Food Policy Research Institute. 2002. Green Revolution: Curse or blessing? Available from: http://www.ifpri.org/sites/default/files/pubs/pubs/ib/ib11.pdf.

Ingale, A.G. and A.N. Chaudhari. 2013. Biogenic synthesis of nanoparticles and potential applications: An eco-friendly approach. J. Nanomed. Nanotechol. 4: 165.

Ion, A.C., I. Ion and A. Culetu. 2010. Carbon-based nanomaterials: Environmental applications. Univ. Politehn. Bucharest 38: 129–132.

Joseph, T. and M. Morrison. 2006. Nanotechnology in agriculture and food institute of nanotechnology. A nanoforum report, retrieved from http://www. nanoforum.org/dateien/temp/nanotechnology.

Kah, M. 2015. Nanopesticides and nanofertilizers: Emerging contaminants or opportunities for risk mitigation? Front. Chem. 3: 64.

Karn, B., T. Kuiken and M. Otto. 2009. Nanotechnology and *in situ* remediation: A review of benefits and potential risks. Environ. Health Persp. 117(12):1823-1831.

Kashyap, P.L., X. Xiang and P. Heiden. 2015. Chitosan nanoparticle-based delivery systems for sustainable agriculture. Int. J. Biolog. Macromol. 77: 36–51.

Katiyar, D., A. Hemantaranjan and B. Singh. 2015. Chitosan as a promising natural compound to enhance potential physiological responses in plant: a review. Indian J. Plant Physiol. 20: 1–9.

Khodakovskaya, M.V., K. de Silva, A.S. Biris, E. Dervishi and H. Villagarcia. 2012. Carbon nanotubes induce growth enhancement of tobacco cells. ACS Nano 6(3): 2128–2135.

Kim, S.W., J.H. Jung, K. Lamsal, Y.S. Kim, J.S. Min and Y.S. Lee. 2012. Antifungal effects of silver nanoparticles (AgNPs) against various plant pathogenic fungi. Mycobiol. 40: 53–58.

Kirchmann, H. and G. Thorvaldsson. 2000. Challenging targets for future agriculture. Eur. J. Agron. 12: 145–161.

Knorr, W., I.C. Prentice, J.I. House and E.A. Holland. 2005. Long-term sensitivity of soil carbon turnover to warming. Nature 433: 298–302.

Lee, J.E., N. Lee, T. Kim, J. Kim and T. Hyeon. 2011. Multifunctional mesoporous silica nanocomposite nanoparticles for theranostic applications. Acc. Chem. Res. 44: 893–902.

Liu, F., L.X. Wen, Z.Z. Li, W. Yu, H.Y. Sun and J.F. Chen. 2006. Porous hollow silica nanoparticles as controlled delivery system for water soluble pesticide. Mat. Res. Bull. 41: 2268–2275.

Liu, B., X.Y. Li, B.L. Li, B.Q. Xu and Y.L. Zhao. 2009. Carbon nanotube-based artificial water channel protein: Membrane perturbation and water transportation. Nano Lett. 9(4): 1386–1394.

Liu, R. and R. Lal. 2015. Potentials of engineered nanoparticles as fertilizers for increasing agronomic productions. Sci. Total Environ. 514: 131–139.

Llop, J., I. Estrela-Lopis, R.F. Ziolo, A. González, J. Fleddermann, M. Dorn, V.G. Vallejo, R. Simon-Vazquez, E. Donath, Z. Mao, C. Gao and S.E. Moya. 2014. Uptake, biological fate, and toxicity of metal oxide nanoparticles. Part. Part. Syst. Charact. 31: 24–35.

Lu, C.M., C.Y. Zhang, J.Q. Wen, G.R. Wu and M.X. Tao. 2002. Research of the effect of nanometer materials on germination and growth enhancement of glycine max and its mechanism. Soybean Sci. 21(3): 168–171.

Mansha, M., A. Qurashi, N. Ullah, F.O. Bakare, I. Khan and Z.H. Yamani. 2016. Synthesis of In2O3/graphene heterostructure and their hydrogen gas sensing properties. Ceram. Int. 42: 11490–11495.

Manzer, H.S. and H.A.W. Mohamed. 2014. Role of nano-SiO$_2$ in germination of tomato (*Lycopersicum esculentum* Mill) seeds. Saudi J. Bio. Sci. 21: 13-17.

Marzbani, P., Y.M. Afrouzi and A. Omidvar. 2015. The effect of nano-zinc oxide on particleboard decay resistance. Maderas Cienc. Tecnol. 17: 63–68.

Maurice, P.A. and M.F. Hochella. 2008. Nanoscale particles and processes: A new dimension in soil science. Adv. Agron. 100: 123–153.

McLamore, E.S., A. Diggs, P. Calvo-Marzal, J. Shi, J.J. Blakeslee, W.A. Peer, A.S. Murphy and D.M. Porterfield. 2010. Non-invasive quantification of endogenous root auxin transport using an integrated flux microsensor technique. Plant J. 63: 1004–1016.

Mukhopadhyay, S.S. 2005. Weathering of soil minerals and distribution of elements: Pedochemical aspects. Clay Res. 24: 183–199.

Mukhopadhyay, S.S. 2014. Nanotechnology in agriculture: Prospects and constraints. Nanotechnol. Sci. Appl. 7: 63–71.

Naderi, M.R. and A. Danesh-Shahraki. 2013. Nanofertilizers and their roles in sustainable agriculture. Intl. J. Agri. Crop Sci. 5(19): 2229–2232.

Nel, A., T. Xia, L. Madler and N. Li. 2006. Toxic potential of materials at the nano level. Sci. 311: 622–627.

[NNCO] National Nanotechnology Coordinating Office 2009. Nanotechnology-enabled sensing. Report of the National Nanotechnology Initiative Workshop, May 5–7, 2009. 42 pg.

Nuruzzaman, M., M.M. Rahman, Y. Liu and R. Naidu. 2016. Nanoencapsulation, nano-guard for pesticides: A new window for safe application. J. Agric. Food Chem. 64: 1447–1483.

Ombodi, A. and M. Saigusa. 2000. Broadcast application versus band application of polyolefin coated fertilizer on green peppers grown on andisol. J. Plant Nutr. 23: 1485–1493.

Park, H.J., S.H. Kim, H.J. Kim and S.H. Choi. 2006. A new composition of nanosized silica-silver for control of various plant diseases. Plant Pathol. J. 22: 25–34.

Patra, J.K. and K.H. Baek. 2017. Antibacterial activity and synergistic antibacterial potential of biosynthesized silver nanoparticles against foodborne pathogenic bacteria along with its anticandidal and antioxidant effects. Front. Microbiol. 8: 167.

Paula, H.C.B., F.M. Sombra, R.F. Cavalcante, F.O.M.S. Abreu and R.C.M. de Paula. 2011. Preparation and characterization of chitosan/cashew gum beads loaded with *Lippia sidoides* essential oil. Mater. Sci. Engg. 31: 173–178.

Prasad, K.S., D. Pathak, A. Patel, P. Dalwadi, R. Prasad, P. Patel and S.K. Kaliaperumal. 2011. Biogenic synthesis of silver nanoparticles using *Nicotiana tobaccum* leaf extract and study of their antibacterial effect. Afr. J. Biotechnol. 9(54): 8122–8130.

Prasad, R., V.S. Swamy and A. Varma. 2012b. Biogenic synthesis of silver nanoparticles from the leaf extract of *Syzygium cumini* (L.) and its antibacterial activity. Int. J. Pharm. Bio Sci. 3(4): 745–752.

Prasad, R. and V.S. Swamy. 2013. Antibacterial activity of silver nanoparticles synthesized by bark extract of *Syzygium cumini*. J. Nanopart. 2013: 1–6, http://dx.doi.org/10.1155/2013/431218.

Prasad, R., V. Kumar and K.S. Prasad. 2014. Nanotechnology in sustainable agriculture: Present concerns and future aspects. Afr. J. Biotechnol. 13(6): 705–713.

Prasad, T.N.V.K.V., P. Sudhakar, Y. Sreenivasulu, P. Latha, V. Munaswamy, K. Raja Reddy, T.S. Sreeprasad, P.R. Sajanlal and T. Pradeep. 2012a. Effect of nanoscale zinc oxide particles on the germination, growth and yield of peanut. J. Plant Nutri. 35(6): 905–927.

Raliya, R., J.C. Tarafdar, S.K. Singh, R. Gautam, K. Gulecha, K. Choudhary, V. Maurino and V. Saharan. 2014a. MgO nanoparticles biosynthesis and its effect on chlorophyll contents in the leaves of clusterbean (*Cyamopsis tetragonoloba* L.). Adv. Sci. Eng. Med. 6: 1–8.

Raliya, R., J.C. Tarafdar, H. Mahawar, R. Kumar, P. Gupta, T. Mathur, R.K. Kaul, P. Kumar, R. Gautam, A. Kaliya, S.K. Singh and H.S. Gehlot. 2014b. ZnO nanoparticles induced exopolysaccharide production by *B. subtilis* strain JCT1 for arid soil applications. Intl. J. Biol. Macromol. 65: 362–368.

Ramacharyulu, P.V.R.K., R. Muhammad, J. Praveen Kumar, G.K. Prasad and P. Mohanty. 2015. Iron phthalocyanine modified mesoporous titania nanoparticles for photocatalytic activity and CO$_2$ capture applications. Phys. Chem. Chem. Phys. 17: 26456–26462.

Ramalingam, C., N. Dasgupta, S. anjan, D. Mundekkad, R. Shanker and A. Kumar. 2015. Nanotechnology in agro-food: From field to plate. Food Res. Int. 69: 381–400.

Rana, S. and P.T. Kalaichelvan. 2013. Ecotoxicity of nanoparticles. ISRN Toxicol. 2013: 574648. doi:10.1155/2013/574648.

Rawal, I. and A. Kaur. 2013. Synthesis of mesoporous polypyrrole nanowires/nanoparticles for ammonia gas sensing application. Sens. Actuators A Phys. 203: 92–102.

Rodell, M., I. Velicogna and J.S. Famiglietti. 2009. Satellite-based estimates of groundwater depletion in India. Nature 460: 999–1002.

Rusonik, I., H. Polat, H. Cohen and D. Meyerstein. 2003. Reaction of methyl radicals with metal powders immersed in aqueous solutions. Eur. J. Inorg. Chem. (23): 4227–4233.

Sagadevan, S. and M. Periasamy. 2014. Recent trends in nanobiosensors and their applications—a review. Rev. Adv. Mater. Sci. 36: 62–69.

Saharan, V., R.V. Kumaraswamy, R.C. Choudhary, S. Kumari, A. Pal, R. Raliya and P. Biswas. 2016. Cu-Chitosan nanoparticle mediated sustainable approach to enhance seedling growth in maize by mobilizing reserved food. J. Agric. Food Chem. 64: 6148–6155.

Scrinis, G. and K. Lyons. 2007. The emerging nano-corporate paradigm: Nanotechnology and the transformation of nature, food and agri-food systems. Int. J. Sociol. Food Agric. 15: 22–44.

Selivanov, V.N. and E.V. Zorin. 2001. Sustained action of ultrafine metal powders on seeds of grain crops. Perspekt. Materialy 4: 66–69.

Shaalan, M., M. Saleh, M. El-Mahdy and M. El-Matbouli. 2016. Recent progress in applications of nanoparticles in fish medicine: A review. Nanomed. Nanotechnol. Biol. Med. 12: 701–710.

Shah, M.A. and A. Towkeer. 2010. Principles of nanosciences and nanotechnology. Narosa Publishing House, New Delhi.

Sharma, V.K., R.A. Yngard and Y. Lin. 2009. Silver nanoparticles: Green synthesis and their antimicrobial activities. Adv. Colloid Interface Sci. 145: 83–96.

Shin, W.K., J. Cho, A.G. Kannan, Y.S. Lee and D.W. Kim. 2016. Cross-linked composite gel polymer electrolyte using mesoporous methacrylate-functionalized SiO$_2$ nanoparticles for lithium-ion polymer batteries. Sci. Rep. 6: 26332.

Shweta, D.K. Tripathi, S. Singh, S. Singh, N.K. Dubey and D.K. Chauhan. 2016. Impact of nanoparticles on photosynthesis: Challenges and opportunities. Mater. Focus 5: 405–411.

Silva, M.S., D.S. Cocenza, R. Grillo, N.F. de Melo, P.S. Tonello, L.C. de Oliveira, D.L. Cassimiro, A.H. Rosa and L.F. Fraceto. 2011. Paraquat-loaded alginate/chitosan nanoparticles: Preparation, characterization and soil sorption studies. J. Hazard. Mater. 190: 366–374.

Singh, S., K. Vishwakarma, S. Singh, S. Sharma, N.K. Dubey, V.K. Singh, S. Liu, D.K. Tripathi and D.K. Chauhan. 2017. Understanding the plant and nanoparticle interface at transcriptomic and proteomic level: A concentric overview. Plant Gene 11: 265–272.

Somasundaran, P., X. Fang, S. Ponnurangam and B. Li. 2010. Nanoparticles: Characteristics, mechanisms and modulation of biotoxicity. Kona Powder and Particle J. 28: 38–49.

Sugunan, A. and J. Dutta. 2008. Nanotechnology Vol. 2: Environmental Aspects (Krug Harald (ed.). Wiley-VCH, Weinheim.

Suman, P.R., V.K. Jain and A. Varma. 2010. Role of nanomaterials in symbiotic fungus growth enhancement. Curr. Sci. 99: 1189–1191.

Susha, V.S., C.R. Chinnamuthu and K. Pandian. 2009. Remediation of herbicide atrazine through metal nanoparticle. *In*: International Conf. Magnetic Materials and their Applications in the 21st Century, October 21–23, 2008, organized by the Magnetic Society of India, National Physical Laboratory, New Delhi.

Swamy, V.S. and R. Prasad. 2012. Green synthesis of silver nanoparticles from the leaf extract of *Santalum album* and its antimicrobial activity. J. Optoelectron. Biomed. Mater. 4(3): 53–59.

Tarafdar, J.C., S. Sharma and R. Raliya. 2013. Nanotechnology: Interdisciplinary science of applications. Afr. J. Biotechnol. 12(3): 219–226.

Teodoro, S., B. Micaela and K.W. David. 2010. Novel use of nano-structured alumina as an insecticide. Pest Manag. Sci. 66(6): 577–579.

Tillman, D., K.G. Cassman, P.A. Matson, R. Naylor and S. Polasky. 2002. Agricultural sustainability and intensive production practices. Nature 418: 671–677.

Tilman, D., J. Knops, D. Wedin and P. Reich. 2002. Plant diversity and composition: Effects on productivity and nutrient dynamics of experimental grasslands. pp. 21–35. *In*: Loreau, M., S. Naeem and P. Inchausti (eds.). Biodiversity and Ecosystem Functioning. Oxford University Press, Oxford.

Torney, F. 2009. Nanoparticle mediated plant transformation. Emerging technologies in plant science research. Interdepartmental Plant Physiology Major Fall Seminar Series. Phys. p. 696.

Trenkel, M.E. 1997. Controlled-release and stabilized fertilizers in agriculture. International Fertilizer Industry Association, Paris.

Tripathi, D.K., Shweta, S. Singh, S. Singh, R. Pandey, V.P. Singh, N.C. Sharma, S.M. Prasad, N.K. Dubey and D.K. Chauhan. 2016a. An overview on manufactured nanoparticles in plants: Uptake, translocation, accumulation and phytotoxicity. Plant Physiol. Biochem. 110: 2–12.

Tripathi, D.K., S. Singh, V.P. Singh, S.M. Prasad, D.K. Chauhan and N.K. Dubey. 2016b. Silicon nanoparticles more efficiently alleviate arsenate toxicity than silicon in maize cultivar and hybrid differing in arsenate tolerance. Front. Environ. Sci. 4: 46. doi:10.3389/fenvs.2016.000 46.

Tripathi, D.K., S. Singh, S. Singh, P.K. Srivastava, V.P. Singh, S. Singh, S.M. Prasad, P.K. Singh, N.K. Dubey, A.C. Pandey and D.K. Chauhan. 2017a. Nitric oxide alleviates silver nanoparticles (AgNps)-induced phytotoxicity in *Pisum sativum* seedlings. Plant Physiol. Biochem. 110: 167–177.

Tripathi, D.K., R.K. Mishra, S. Singh, S. Singh, K. Vishwakarma, S. Sharma, V.P. Singh, P.K. Singh, S.M. Prasad, N.K. Dubey, A.C. Pandey, S. Sahi and D.K. Chauhan. 2017b. Nitric oxide ameliorates zinc oxide nanoparticles phytotoxicity in wheat seedlings: Implication of the ascorbate-glutathione cycle. Front. Plant Sci. 8:1. doi: 10.3389/fpls.2017.00001.

Tripathi, D.K., A. Tripathi, Shweta, S. Singh, Y. Singh, K. Vishwakarma, G. Yadav, G., S. Sharma, V.K. Singh, R.K. Mishra, R.G. Upadhyay, N.K. Dubey, Y. Lee and D.K. Chauhan. 2017c. Uptake, accumulation and toxicity of silver nanoparticle in autotrophic plants, and heterotrophic microbes: A concentric review. Front. Microbiol. 8: 07. doi:10.3389/fmicb.2017. 00007.

Ullah, R. and J. Dutta. 2008. Photocatalytic degradation of organic dyes with manganese-doped ZnO nanoparticles. J. Hazard Mater. 156: 194.

Ullah, H., I. Khan, Z.H. Yamani and A. Qurashi. 2017. Sonochemical-driven ultrafast facile synthesis of SnO_2 nanoparticles: Growth mechanism structural electrical and hydrogen gas sensing properties. Ultrason. Sonochem. 34: 484–490.

Ulriches, C., I. Mewis and A. Goswami. 2005. Crop diversification aiming nutritional security in West Bengal: Biotechnology of stinging capsules in nature's water-blooms. Ann. Tech. Issue of State Agri. Technologists Service Association. 1–18.

[UN] United Nations. 2015. World Population Projected to Reach 9.7 Billion by 2050; UN: New York; http://www.un.org/en/development/desa/news/population/2015-report.html.

Vannini, C., G. Domingo, E. Onelli, F. De Mattia, I. Bruni, M. Marsoni and M. Bracale. 2014. Phytotoxic and genotoxic effects of silver nanoparticles exposure on germinating wheat seedlings. J. Plant Physiol. 171: 1142–1148.

Vidhyalakshmi, R., R. Bhakyaraj and R.S. Subhasree. 2009. Encapsulation the future of probiotics–A review. Adv. Biol. Res. 3(3-4): 96–103.

Wang, L., Z. Li, G. Zhang, J. Dong and J. Eastoe. 2007. Oil-in-water nanoemulsions for pesticide formulations. J. Colloid Interface Sci. 314: 230–235.

Waychunas, G.A., C.S. Kim and J.A. Banfield. 2005. Nanoparticulate iron oxide minerals in soils and sediments: Unique properties and contaminant scavenging mechanisms. J. Nanopart Res. 7: 409–433.

Wu, L., M. Liu and R. Liang. 2008. Preparation and properties of a double-coated slow-release NPK compound fertilizer with superabsorbent and water-retention. Bioresour. Technol. 99: 547–554.

Yamanka, M., K. Hara and J. Kudo. 2005. Bactericidal actions of silver ions solution on *Escherichia coli* studying by energy filtering transmission electron microscopy and proteomic analysis. Appl. Environ. Microbiol. 71: 7589–7593.

Yang, F.L., X.G. Li, F. Zhu and C.L. Lei. 2009. Structural characterization of nanoparticles loaded with garlic essential oil and their insecticidal activity against *Tribolium castaneum* (Herbst) (Coleoptera: Tenebrionidae). J. Agric. Food Chem. 57(21): 10156–10162.

Zhan, J., P.H. Thrall, J. Papaïx, L. Xie and J.J. Burdon. 2015. Playing on a pathogen's weakness: Using evolution to guide sustainable plant disease control strategies. Ann. Rev. Phytopathol. 53: 19–43.

Zhang, Q., L. Han, H. Jing, D.A. Blom, Y. Lin, H.L. Xin and H. Wang. 2016. Facet control of gold nanorods. ACS Nano 10: 2960–2974.

CHAPTER 5

Agri-Applications of Nano-Scale Micronutrients

Prospects for Plant Growth Promotion and Use-Efficient Micronutrient Fortification

Anu Kalia[1],* *and Harleen Kaur*[2]

1. Introduction

Fertilizers are synthetic/natural compounds formulated in appropriate concentrations and combinations, which furnish the essential nutrient elements to plants on application to achieve improved growth and yield of crop plants (Behera and Panda 2009). Fertilizers are required to achieve higher food production targets, 'more crop per unit agrable land', to satiate the burgeoning human population. As intensive crop cultivation depletes the capability of the arable land to sustain the optimum nutrient contents for maintenance of appropriate plant growth due to voracious mining, uptake and exhaustion of essential nutrients by the previous crop (Marschner 2012, Rietra et al. 2015), therefore, nutrients have to be replenished instantaneously as fertilizers. However, recurrent fertilization or, more precisely, the fertilizer misuse and abuse can lead to emergence of numerous negative impacts, such as decreased fertilizer response ratios due to development of non-responsive soils, episodes of multi-micronutrient (MN) deficiencies in soils besides weathering of soil minerals and development of ailing soils incapable of sustaining optimum fertility standards

[1] Electron Microscopy and Nanoscience Laboratory, Department of Soil Science, College of Agriculture.
[2] Department of Microbiology, College of Basic Sciences and Humanities, Punjab Agricultural University, Ludhiana, Punjab, India-141004.
* Corresponding author: kaliaanu@pau.edu, kaliaanu@gmail.com

(Bindraban et al. 2015). Thus, strenuous R&D efforts for formulation of balanced nutrient fertilization regimes, comprised of micronutrients and secondaries alongwith the macronutrients in crop, variety, soil, and agroecosystem specific manner, have to be devised (Joy et al. 2015, Dimpka and Bindraban 2016). Likewise, endeavors in next-gen innovation-driven technology 'nanotechnology' are required in order to address the pitfalls of the conventional fertilizers. This manuscript will explore the benefits of MN fertilization for enhancing the crop productivity, and produce quality with particular relevance to nano-MN fertilizers.

Micronutrients are the mineral elements required in very low or trace quantities by the plants for proper growth and development. These elements play a pivotal role in physiological and biochemical processes as these function as co-factors or conjugating metal central atom of several enzymes and primary/secondary biological macromolecules. The essential micronutrients include boron (B), copper (Cu), iron (Fe), manganese (Mn), molybdenum (Mo), nickel (Ni) and zinc (Zn). Other micronutrients, such as sodium (Na), selenium (Se) and silicon (Si), are considered non-essential but have been considered important for plant growth (Dimkpa and Bindraban 2016).

2. Why Crop Plants Exhibit Micronutrient Demands?

The agrable land under intensive cultivation system is likely to suffer substantial loss of nutrients as the growing crop forges for the necessary macro-/micro-nutrients. This biological mining preferentially depletes the MNs from and around the root zone as the yield-limiting macro-nutrients, such as nitrogen, phosphorus and potassium, are replenished in the soil through application of synthetic fertilizers like triple fertilizer NPK (Bindraban et al. 2015). Consequently, the deficiency of these MN elements becomes a limiting factor that may even affect the use efficiency of the macro-nutrients and other inputs, such as water, causing deficiency diseases epitomized as abnormal or reduced growth besides reduction in the yield attributing characters and, thus, yield and productivity.

2.1 Role of Micronutrients in Plants: How Nano-Scaling of Nutrients Improves their Role?

Micronutrients are present in the earth crust and are industrially processed for their use as fertilizers. Micronutrient deficiency in soil is observed globally, although the extent of a particular nutrient deficiency may vary (Voortman and Bindraban 2015). These essential trace elements exhibit intricate and complex interactions with a myriad of soil abiotic and biotic components, debarring their relative bioavailability. Compounding these problems are the aberrant climate changes that lead to alteration in edaphic factors and components, thereby further limiting the plant-available forms of these nutrients in soil (Wang et al. 2016b). Micronutrients are required by the crop plants for maintenance of their basic physiological and metabolic properties (Table 1), for alleviation of elevated biotic or abiotic stresses imparted during different growth conditions of the plant, and for improving the nutritional quality of the produce.

2.1.1 Agronomic Relevance

The MN demand by plants varies depending upon the nutritional requirements of the plant. In most of the higher plants, the MNs are accumulated in the following order: Mn > Fe > Zn > B > Cu > Mo (Fageria et al. 2002). Moreover, the uptake also depends upon several factors, such as the plant nutrient demand, root architecture and other root properties, such as root hair density (Wang et al. 2016a), lateral root density (contribute significantly for acquisition of phosphorus, and MNs like Mn, Zn, and Cu; Liu et al. 2013) and type of root exudates (Dakora and Phillips 2002, Shen et al. 2013), activity of various soil enzymes and microbial associations (Tkacz and Poole 2015, Krithika and Balachandar 2016, Kamran et al. 2017) (Table 2). Interestingly, the bioavailable concentrations of MN(s) can invariably affect these factors, particularly the plant root architecture, due to *in planta* production of the indole acetic acid (Shahzad and Amtmann 2017).

Application of MN fertilizers can help to enhance crop growth and yield, particularly in nutrient-deficient soils. These types of soils are developed due to biological mining of the MNs preferentially under intensive cultivation conditions vis-à-vis application of only major nutrients, i.e., NPK (Dimpka and Bindraban 2016). The problem becomes recalcitrant due to low use efficiency (approximately < 5%) of the soil-applied MN fertilizers (Monreal et al. 2016). The crops growing on these soils also suffer from low use efficiency of applied NPK, thereby limiting the anticipated growth and yield benefits. Moreover, the success and, thus, the benefits imparted by particular MN fertilizer varies with the application route; for instance, foliar application of Zn, Fe, Mn fertilizers while foliar/soil application of B and Cu fertilizers result in better uptake and growth enhancement (Dimpka and Bindraban 2016) (Fig. 1). The seed germination, and root and shoot growth can be further enhanced by application of MN nanofertilizers. The nano-particulate size endures these fertilizers to be taken up as crystalline structures which are further dissolved in the cytoplasm of the plant cell in order to ensure slow and sustained release of the respective cation/ anion. Thus, the stimulatory effect is dichotomized; (i) due to the nano-size dependent properties and (ii) due to availability of the micronutrient element in cationic/anionic form. Physiologically, the enhancement in the *in planta* indole acetic acid (IAA) levels can be speculated to cause higher root growth. The nano-particulate size dimensions are anticipated to improve micronutrient content in the produce and grain. Pandey et al. (2010) found that ZnO NPs can enhance seed germination and root growth of common gram *Cicer arietinum* (Chickpea) by increasing the level of IAA in roots.

Table 1. Plant-available forms of major micronutrients, uptake proteins and prime functions performed in crop plants.

Micronutrient Element	Symbol	Native soil Mineral Form	Plant-Available Form	Uptake Proteins Involved	Concentration in Plants (mg kg^{-1})	Function Played	References
Boron	B	Borax, Kernite, Tournamile, Colemanite	BO_3^{3-} (Borate ion) Uncharged boric acid [$B(OH)_3$]	High affinity channel transport proteins (BOR1, and aquaporin NIPS; 1) [Miwa et al. 2010]	0.2–800	Involved in cell wall synthesis, protein synthesis, respiration, proton pumping causing hyperpolarization of membrane potential	Blevins and Lukaszewski 1998, Hansch and Mendel 2009
Chlorine	Cl	Potassium chloride	Cl^-	Both passive (apoplastic) and active symplast movement via trans-membrane ion transporters (symporters/antiporters), Membrane ATPases [White and Broadely 2001, Li et al. 2017]	10–80,000	Component of oxygen evolving complex, stimulation of ATPase at tonoplast	Churchill and Sze 1984, Kusunoki 2007
Copper	Cu	Malachite, Cuprite, Azurite, Chalcopyrite	Cu^+, Cu^{2+}, CuO	Cation transporter (Cu transporter COPT/Ctr-like protein family particularly COPT6) [Printz et al. 2016]	2–50	Required for photosynthesis, oxidative stress defence, mitochondrial respiration; Exists as copper chaperons	Huffman and O'Halloran 2001
Iron	Fe	Siderite, Hematite, Magnetite, Pyrite	Fe^{2+}, FeO	Cation transporter (Iron-Regulated Transporter 1) through reducing and phytosiderophore-chelating strategies [Connorton et al. 2017]	20–600	Involved in respiration, photosynthesis (80% iron is present in chloroplasts), phytohormone synthesis	Hansch and Mendel 2009

Manganese	Mn	Manganite, Rhodocrosite, Braunite	Mn^{2+}	Cation transporter (NRAMP transmembrane channel proteins) [Socha and Guerinot 2014, Shao et al. 2017]	10–600	Co-factor in enzymes involved in redox, decarboxylation and hydrolytic reactions	Grusak 2001
Molybdenum	Mo	Ilsemanite, Powellite, Molybdenite	MoO_4^{2-} (molybdate ion)	S transporter [Fitzpatrick et al. 2008], specific transporter proteins MOT1 and MOT2 [Gasber et al. 2011]	0.1–10	Component of nitrogenase, involved in sulfur metabolism and phytohormone synthesis	Schwarz and Mendel 2006, Nie et al. 2016
Nickel	Ni	Pentlandite, Awaruite, Cohenite	Ni^{2+}	Cation transporter (IRT1 transporter) [Nishida et al. 2011]	0.05–5	Co-factor in enzyme urease; Exists as nickel chaperon	Bai et al. 2006
Zinc	Zn	Smithsonite, Hemimorphite	Zn^{2+}	Cation transporter (transporters ZIP (ZRT, IRT-like protein-IRT1) family and intracellular high-affinity binding sites) [Bouain et al. 2014]	10–250	Co-factor of enzymes involved in protein synthesis; Involved in maintaining structural integrity of membranes	Hansch and Mendel 2009

Table 2. Micronutrient nanofertilizers affecting the soil microbial diversity and alterations in the functioning of the specific microbial groups involved in biogeochemical cycling.

Micronutrient	Nanoparticle Used, Dosage	Soil Community Affected	Physiological Effect	References
Copper	CuO (1% w/w to dry soil)	Soil bacteria	Decrease in bacterial community composition	Ben-Moshe et al. 2013
	CuO (1, 10 and 100 mg kg^{-1})	Soil microbial communities	Increase in cellobiohydrolase, β-1,4-xylosidase and β-1,4-glucosidase enzyme activities at 10 mg kg^{-1} concentrations after 30 days of exposure	Asadishad et al. 2018
Iron	Fe$_3$O$_4$ (100 and 200 ppm)		Increase in arbuscular mycorrhizal colonization with increase in conc., higher nodule colonization	Burke et al. 2015
Zinc	ZnO (1000 mg kg^{-1})	Bacterial and fungal communities in date palm leaf litter amended soil	Reduction in number of colonies	Rashid et al. 2017
	ZnO (100–1600 µg ml^{-1})	*P. aeruginosa, B. amyloliquefaciens* and *P. fluorescens* strains	Decrease in IAA production, siderophore production and phosphate solubilization	Haris and Ahmad 2017
Molybdenum	Mo (8 mg l^{-1})	Rhizospheric communities in chickpea	Increase in number of nitrifiers, oligotrophs, nitrogen utilizers, *Azotobacter*, actinomycetes and phosphorus mobilizers at emerging stage of plant as compared to *Bradyrhizobium* inoculation	Taran et al. 2014

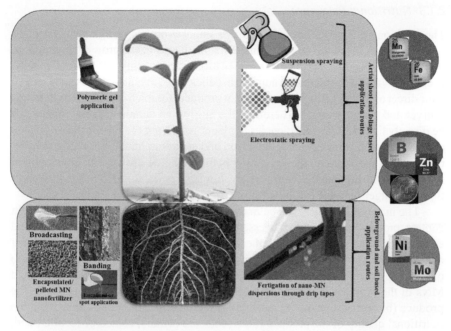

Fig. 1. Micronutrient nanofertilizers and their desirable or preferred application routes.

2.1.2 Physiological Relevance

Micronutrients have a demonstrable impact on the physiology of the crop plants because they act as co-factor/co-enzyme components of an array of enzymes involved in metabolic cascades; particular mention is for the anti-oxidant and stress-related enzyme systems (Hajiboland 2012). Micronutrients also improve the ability of a plant to combat a range of stress conditions due to induction of stress tolerance to diverse biotic (Dordas 2008, Cacique et al. 2012, Veresogloua et al. 2013, Kaur et al. 2016) and abiotic factors (Rahimizadeh et al. 2007, Waraich et al. 2012, Ahanger et al. 2016). A plant with better nutritional status and metabolic activities can be expected to possess greater capacity to adapt to stress conditions (Moralez-Diaz et al. 2017). The nanofertilizers containing nutrient elements such as Zn, Cu, Fe, Mn, Se and others, exhibit an enhanced induction of the oxidative stress in microorganisms, aquatic and terrestrial plants which may also impart resistance to abiotic (Elmer and White 2016) and biotic (Boregowda et al. 2017) stress. However, in the case of terrestrial plants, if the oxidative stress level does not exceed the toxic threshold level responsible for cell death, it leads to induction of a defense response phenomenon that results in expression of resistance genes and accumulation of enzymatic and non-enzymatic antioxidants (Van Aken 2015).

2.1.3 Nutritional Relevance

The nutritional quality of the produce is as essential as obtaining higher yield per unit agrable land to fulfill the food and nutritional demands of the growing population. Therefore, the focus needs to be on maintenance or improvement of the food quality, keeping in view of the increased nutritional deficiencies in humans and livestock and their direct or indirect dependence on plant products for micronutrient requirements (Oliver and Gregory 2015). Therefore, the nutritional deficiencies in humans can be directly linked to MN deficiencies in plants, rendering the fortification of plants a necessary step. Among these, zinc, iron and selenium have been recognized by the WHO as the essential dietary nutrients for the better growth and development of animal and human life (http://www.lbhi.is/sites/default/files/gogn/vidhengi/thjonusta/utgefid_efni/RitLbhi/Rit_LbhI_nr_3.pdf).

The nutritional quality can be manifested as improvement in the total protein content, protein quality (amount of essential amino acids) and antioxidant compound enrichment, such as carotenoids, phenolics and MN fortification. The nutritional quality can be improved by application of MN fertilizers at specific growth stage(s) of a crop plant. This may lead to better partitioning and accumulation of applied MNs in the produce or the grain, thereby enhancing the nutritive quality of the produce (biofortification). The nutrients in nano-form will help in improving the nutritional quality of the produce if the seeds are primed, coated or fortified with the nanoparticles of these elements (Subbiah et al. 2016). Few examples to quote are significant increase in the iron content in ginger plants by application of ferric oxide nanoparticles (Siva and Benita 2016). Similarly, the protein content and quality can also be affected or improved by application of nano-MN fertilizers. Application of nano-Fe improved the seed protein content by 2% compared to Fe in control black-eyed pea (Delfani et al. 2015).

2.2 Role of the Growth Matrix 'Soil' in Micronutrient Availability

The soil is a dynamic system comprised of the non-living edaphic and living biotic components. An interplay among these components is helpful in sustaining the physical structure, chemical properties and biological fertility of the soil. Micronutrients on application exhibit interaction with these soil components which can directly or indirectly affect their bioavailability.

2.2.1 Soil Micronutrient Deficiency: How far Does it Affects Crop Growth?

The MN trace requirements are indispensable for ensuring proper growth and development of the crop plants and, thus, their deficiencies may impair it (Table 3). The micronutrients must be present in adequate concentrations in the soil to be taken up by the plant, since higher concentrations are as detrimental as the lower amounts (Sikka et al. 2018 in press). Besides the MNs should be present in bioavailable form for their easy uptake. As discussed in the previous section, the availability

Table 3. Deficiency symptoms observed in crop plants due to lower concentrations of micronutrients in soil.

Micronutrient	Deficiency Symptom
Boron (B)	Black necrosis of young leaves
Copper (Cu)	Necrotic spots arising from tips
Iron (Fe)	Intervenous chlorosis
Magnesium (Mg)	Chlorosis
Manganese (Mn)	Intervenous chlorosis
Molybdenum (Mo)	Necrosis, premature flower abscission
Nickel (Ni)	Urea accumulation in leaves
Zinc (Zn)	Rosetting

of these nutrients will depend upon soil and root structure, soil pH, quantity of organic matter, cation exchange capacity and presence of microbial associations (Solanki et al. 2015). The MN impoverished crop plants remain stunted, exhibit chlorosis and reduced photosynthesis related reactions, may not bear flowers or fruits and may even die under severe deficiency conditions (Uchida 2000, Sinclair and Edwards 2008).

2.2.2 Dynamics of Applied Nano-Micronutrients in Soil

The soil-applied MN fertilizers undergo a variety of interactions with the physical, chemical and biological components of the dynamic soil system undergoing recurrent redox alterations (oxido-reduction events) and, hence, may be transformed into plant unavailable forms. Similarly, the nano-scale MNs are also expected to interact with the soil matrix in a myriad of manners. However, the diversity in morphology (shape, size, size distribution, dimensions), crystallinity (porous or non-porous), purity, dose of application (lower or higher) and surface properties (surface charge, chemistry, redox potential) of the nano-MN material can substantially affect the behavior of applied MN nanoparticles (NPs) in soil (Khare et al. 2015). In a similar manner, these properties may also affect the toxicity of these NPs. The soil system being the major sink for the applied MN NPs may also accomplish mitigation of NP-toxicity. The soil components, primarily the humic acid and soil pore water, can act as buffers that dilute the deleterious antimicrobial action of the applied NPs on soil/plant beneficial microflora (Calder et al. 2012). Moreover, the possibilities of NP transformations and subsequent generation of toxic intermediates may further enhance the ROS-dependent processes in microbes and plants, thereby attenuating the soil NP-toxicity mitigation phenomena.

The NPs may either undergo aggregation or agglomeration (due to strong and weak van der Waal interactions among the component NPs in the formulation) (Jiang et al. 2009), may exhibit adsorption to living or non-living surfaces (such as microbial, plant and soil micro and macrofaunal cells and soil colloids, minerals and organic components) and may evince dissolution to generate the respective mono or divalent cation/anion (Karimi and Fard 2017). Consequently, this may alter the pH and release of nutrients through the soil CEC (Conway and Keller 2016). The aggregation/agglomeration of applied MN NPs on interaction with soil is largely governed by NP characteristics as well as by physical forces, such as NP random motion, gravitational force and fluid motion (Conway and Keller 2016). Formation of micrometric aggregates, such as for ZnO and CuO MN-NF, resulted in their decreased mobility, leading to decreased reactivity with the soil components (Dimpka et al. 2013). Presence or absence of surface coatings is also an important determining factor in the fate of nanoparticles in soil. Uncoated metal nanoparticles have charged surfaces and can, therefore, participate in redox reactions with metal ions and ligands present in the soil (Tourinho et al. 2012). Thus, the presence of surface coatings significantly affects the dynamics of nanoparticles in soil. Soil humic and fluvic acids can get adsorbed on NP surface, thereby altering its stability (Ghosh et al. 2008). Absence of the surface coatings may also facilitate dissolution of the ions into the soil matrix. The dissolution of micronutrient nanofertilizers may contribute to toxicity. Further, these NPs may exhibit bioaccumulation in soil micro, meso and macro-flora and fauna, which may lead to biomagnifications by the trophic transfer. The negatively-charged cuticle of the soil nematodes may attract certain NPs (Wang et al. 2009). Exposure of Cu NPs (size 80 nm) decreased the energy reservoir compounds in *Enchytraeus albidus* after an exposure of three weeks in field soil (Amorim et al. 2012), while Zn NPs increased the catalase and superoxide dismutase activities in *E. foetida* (Hu et al. 2010).

Application of nano-MN fertilizers can alter the C and N-mineralization, for instance, the zinc oxide NPs significantly reduced the carbon mineralization (CO_2 emissions and dissolved organic carbon) and also affected nitrogen mineralization in *Phoenix dactylifera* leaf litter-amended soil (Rashid et al. 2017a). Alternatively, several studies advocate that metal/metal oxide NPs do not affect soil properties, such as soil organic matter and microbial biomass carbon content, pH, mineral concentration and available nutrients (Carbone et al. 2014). However, these can alter the bioavailability of nutrients and/or toxic compounds in the soil (Rashid et al. 2017a,b). Thus, the contradictory interactions between NPs and soil limit the large-scale use of MN nanofertilizers in agriculture.

3. How to Prepare Nano-Scale Micronutrient Fertilizers?

The characteristic properties exhibited by nanofertilizers are attributed to their size. Thus, the appropriate synthesis of nano-scale metal/metal oxides is of prime importance and cannot be overlooked. There are two approaches employed for the synthesis of nano-scale micronutrient fertilizers: Bottom-up and Top-down. The bottom-up approach employs atomic or molecular scale components in order to assemble nanoparticles via natural processes or externally applied force, whereas the top-down approach uses refining of larger molecules to attain particles of nano-size. The most common bottom-up fabrication techniques are the sol-gel synthesis, co-precipitation, surfactant assisted microemulsion, physical/chemical vapour deposition and laser pyrolysis. However, the top-down approaches include beam-based or dry (ion beam milling and reactive ion etching) or chemical etching (Dariao and Jin 2012) techniques. The spontaneous or templated assembly of metal or metal oxide nanomaterials can also be carried out by biological or green synthesis techniques. These involve the cell or cell extracts of diverse organisms such as plants (Maruthupandy et al. 2017), bacteria (Revati and Pandey 2011, Jeevanandam et al. 2016, Kaur and Kalia 2016), algae (Fawcett et al. 2017), and fungi (Zielonka and Klimek-Ochab 2017) for the synthesis of the nanomaterials. The biological entities, such as proteins (both structural and enzymatic), polyphenols and other phytochemicals, through their reductive capacities, can transform soluble metal ions into nanoparticles (Makarov et al. 2014). The nano-scale materials thus derived exhibit the phenomena of protein corona formation (Teske and Detweiler 2015). The lack of uniformity in the shape and size of the nanoparticles synthesized by this method is a limitation for its industrial scale application.

4. Conventional vs. Nano-Scale Micronutrients

4.1 Micronutrient Use-Efficiency

Conventional fertilizers suffer from substantial low use efficiency issues. Applied MNs can undergo rapid reaction with organic and inorganic compounds (phosphates and carbonates) and can either form precipitates or can interact with clay colloids, thus, rendering themselves unavailable to the crop (Marschner 2012). The use of nanofertilizers is anticipated to circumvent as well as effectively address these serious low use efficiency (UE) issues of the conventional fertilizers so as to reduce losses and ensure environment safety. Moreover, the applied nutrient UE stays very high for nanofertilizers, as equivalent or higher plant growth and yield may occur at doses substantially lower than the same nutrient supplied in its bulk form (Duhan et al. 2017). Therefore, the benefits of the use of nano-scale micronutrient fertilizers can ensure increased bioavailability, slow and controlled release of the micronutrients, longer lifetime, and reduced losses of nutrients in comparison to the conventional fertilizers (Subramanian and Rahale 2013).

4.2 Uptake, Partitioning, Transportation and Accumulation

The main distinguishing feature of nano-scale MN fertilizer is the mode of uptake and accumulation by plants. Conventional MN fertilizers are often added to NPK fertilizers and are taken up by the plants, alongwith the latter, on soil application or by fertigation. However, the uptake and translocation of nano-MN particles varies according to the plant type, growth conditions, nature of NPs and their mode of delivery (Solanki et al. 2015). On foliar or root application, the NPs enter plant cells depending upon their relative smaller size with respect to the natural pores, or via transmembrane protein channels such as aquaporins, ion channels, specialized membrane transporters and by process of endocytosis (Nair et al. 2010). Following the entry into the cell, these internalized NPs may follow an apoplastic or symplastic route for translocation; the former is non-living pathway allowing the transport of NPs through the cell wall, intracellular free spaces, while the latter includes their journey from within the protoplast via natural cell junctions, 'plasmodesmata' (Wu et al. 2017). Through roots, the NPs can follow both the apoplastic as well as the symplastic route (Deng et al. 2014). On foliar uptake, the NPs enter the plant cells through natural openings on the leaf and aerial stem tissue, such as stomata, hydathodes, lenticels and other pores (Hong et al. 2014). Being primarily immobile, MN NPs can offer better uptake and movement within the plant body and therefore, better accumulation in the shoot and grain parts on maturity.

5. Agri-Applications of Nano-Micronutrients: Gain Begets Profit

5.1 Interaction of Nano-Micronutrients with Plant Tissues

Understanding of the interactions between plants and NPs applied as nutrient fertilizers is of paramount importance from the perspective of crop growth and yield enhancements. Recent literature is flooded with researches emphasizing the use of nanofertilizers for plant growth and biomass enhancement (Subbiah et al. 2016, Burke et al. 2015, Disfani et al. 2016, Mahajan et al. 2011, Mahawar et al. 2017). However, several reports also highlight the harmful impact of applied NPs on plant physiology and functioning (Rico et al. 2014, Majumdar et al. 2015). These contrasting effects can be attributed to the chemical nature of the NPs and their dose for a particular crop which is one of the most important determining factors for plant nutrition.

The interactions of nano-MNs with plants depend upon their size, shape, chemical nature, crystallography, topography and hydrophobicity/hydrophilicity of the former, out of which size and surface area are more important (Wang et al. 2016). Smaller particle size is related to their higher accumulation attributing to their increased toxic nature (Slomberg and Schoenfisch 2012). The NPs can induce oxidative stress in plants by increasing the production of reactive oxygen species (ROS). Application of MN NPs may lead to changes in morphology, physiology and genetics of the plant (Duhan et al. 2017). Various studies have demonstrated the positive effects of NF application on crop plants (Table 4).

5.2 Types of Micro-Nutrient Nanofertilizers

Boron is an important plant micronutrient which is known to play an important role in the following activities: Cell elongation, cell wall formation, indole acetic acid metabolism, translocation of sugars and carbohydrate metabolism (Gupta and Solanki 2013, Da Rocha Pinho et al. 2015). It is taken up by the plants in form of boric acid and depends upon the pH and boron concentration in the soil. The foliar application of nano-boron fertilizer enhanced the leaf boron content and increased

Table 4. Impact of application of micronutrient nanofertilizers in crop plants.

Micronutrient Nanofertilizer	Mode of Application	Plant Species	Effect	References
Iron	Foliar application (0.002%)	Borago (*Borago officinalis*)	Increase in number of seeds and plant biomass as compared to nano-urea application; Decrease in oil yield	Mahmoodi et al. 2017
Carbon bound iron oxide nanoparticles	Soil application (20 mg in Hoagland solution having 5 μM Fe)	Rice (*Oryza sativa*)	Increase in plant biomass, photosynthetic activity, iron accumulation in roots and leaves under iron deficient and calcium stressed conditions; Reduction in calcium accumulation by plant	Sebastian et al. 2017
Nano-boron	Foliar application (0.025%, 0.05%, 0.1%)	Zaghloul date palm	Increase in vegetative growth characteristics, chlorophyll content, nutrient content and yield; Further increase when used in combination with wheat seed sprout extract	Refaai 2014
Copper oxide nanoparticles	Foliar application (50,100 and 200 mg l^{-1})	Cucumber (*Cucumis sativa*)	Decrease in copper content of fruit at all concentrations as compared to control; Decrease in quantity of fruit as compared to bulk counterpart	Hong et al. 2015
	Soil amendment (62.5, 125, 250 and 500 mg kg^{-1})	Bell pepper (*Capsicum annum*)	No significant effect on chlorophyll content as compared to bulk counterpart; Increase in stomatal conductance at 62.5 and 125 mg kg^{-1} concentrations; No significant increase in Cu uptake as compared to bulk CuO	Rawat et al. 2018
Manganese oxide nanoparticles	Seed treatment (0.25, 0.5, 5, 10 and 50 ppm)	Lettuce (*Lactuca sativa*)	Increase in germination percentage at concentration 0.25–5 ppm; Increase in root length at all concentrations	Liu et al. 2016
Zinc oxide nanoparticles	Soil amended (250, 500 and 750 mg kg^{-1})	Alfalfa	Reduction in root biomass; Increase in seed germination at 750 mg kg^{-1} as compared to bulk ZnO	Bandyopadhyay et al. 2015

the number of fruits per tree (yield), leading to formation of fruits with higher total soluble solids, titratable acidity, and juice pH (fruit quality) in pomegranate (Davarpanah et al. 2016).

Copper can exist in its monovalent (Cu^+) or divalent (Cu^{2+}) form. This micronutrient is component of various metabolic compounds, such as plastocyanin, photosystem I, cytochrome oxidase and Cu-Zn superoxide dismutase. Initial studies of the effect of copper nanofertilizer on wheat showed inhibition of seedling growth (Lee et al. 2008). However, in the leafy-green vegetable, lettuce, Cu NP exposure enhanced the root and shoot lengths (Shah and Belozerova 2009). Another study on *Cicer arietinum* showed enhanced seed germination rates, vegetative growth, total chlorophyll and protein contents on exposure to copper NP-carbon nanofiber (Cu-CNM) composite material that ensured the slow release of the micronutrient Cu from the CNM matrix (Ashfaq et al. 2016).

Iron (Fe) is an important component of several proteins involved in electron transport chains and functioning as anti-oxidant enzymes, such as the heme proteins (cytochromes and peroxidases), iron-sulfur proteins (ferrodoxin and superoxide dismutase) and iron-containing enzymes (lipoxygenase). The insolubility of the ferric (Fe^{3+}) ionic form of iron is largely responsible for its scarcity in the soil solution, thus culminating to its very low plant-available form for uptake by the plant despite its abundance in various soils. In barley, application of nano Fe/SiO_2 fertilizer at concentrations of 5, 15 and 25 mg kg^{-1} led to increase in shoot length over the conventional Fe/SiO_2 fertilizer (Disfani et al. 2016). However, in the case of maize, enhanced shoot growth could only be observed at lower concentrations while it decreased at higher amount (25 mg kg^{-1}) as compared to conventional Fe/SiO_2 fertilizer. Continuous exposure of rice seeds to FeO nanoparticles (< 50 nm) at 100 ppm for 10 days helped to enhance the vegetative growth (shoot and root length and plant biomass), while at a higher concentration of 400 ppm, the FeO NPs led to an increase in total chlorophyll, total protein content, peroxidase and catalase activities in the leaves of the test plants (Mankad et al. 2017).

Manganese, after iron, is the second most important micronutrient needed for plant growth. Nano-chelated manganese decreased the number of primary branches in chick pea but increased the number of secondary branches and pod length (Janmohammadi et al. 2017). Manganese nanoparticles showed higher percent germination in lettuce (*Lactuca sativa*) plants at concentrations of 0.25, 0.5 and 5 ppm as compared to conventional MnO particles (Liu et al. 2016), whereas root length was greatly increased upto concentrations of 50 ppm (40.7 mm at 50 ppm and 24.2 mm in control).

Molybdenum (Mo) is present as co-factor in some of the plant enzymes, such as nitrogenase, nitrate reductase and xanthine oxidase, which shows its important role in nitrogen metabolism. Hence, undernutrition of Mo may lead to nitrogen deficiency in plants. Foliar application of nanochelated form of molybdenum (1,2 and 3 gL^{-1}) in peanut plants increased the number of seeds per plant, number of pods per plant, number of lateral branches and biological yield with the increase in concentration as compared to the control (Manjili et al. 2014).

Nickel is essential for the activity of urease enzyme. Conventional fertilizers prepared from rock phosphate have nickel in amounts ranging from 16.8 to 50.4 mg kg^{-1} (Raven and Loeppert 1997). Bioavailability of nickel is reduced due to its strong affinity towards divalent cations present in the soil. Deficiency of nickel may result in accumulation of urea in leaves. Effect of application of uncoated and citrate coated Ni(OH)$_2$ nanoparticles was studied on mesquite (*Prosopis* sp.) plants. It was found that application of uncoated Ni(OH)$_2$ nanoparticles at concentration of 0.1 g increased the root length (Parsons et al. 2010). Seed treatment of *Coriandrum sativum* with nickel nanoparticles showed an increase in shoot weight at 40 ppm concentration and increase in root weight at 20 ppm concentration (Miri et al. 2017).

The amphoteric nature of zinc renders its bioavailability to the growing crop plants to be largely governed by the pH of the soil/system. The pulses and vegetable crops have been voracious feeders of the nutrients from soil and, thus, most of the studies regarding the nano-Zn application have been performed in these crops. Application of nano-chelated zinc via soil and foliar spray significantly increased plant height, canopy width, number of primary and secondary branches and pod length and weight over the control in *Cicer arietinum* crop (Janmohammadi et al. 2017). Positive results in relation to plant biomass, chlorophyll and total protein content were obtained when tomato (*Solanum lycopersicum*) plants were subjected to foliar spray of ZnO nanoparticles (Raliya et al. 2015). Lower concentrations of ZnO nanofertilizer (1.5 ppm) improved plant growth in chick pea (Burman et al. 2013). However, the application of ZnO NPs in cereal crops holds the maximum challenge as well as benefit. Janmohammadi et al. (2016a) have reported that application of nano-chelated Zn enhanced the vegetative, photosynthetic and yield-attributing characters in maize under deficit irrigation (50% field capacity). Therefore, they have recommended the application of Zn nanofertilizers in order to improve maize productivity and yield in semi-arid regions alongwith substantial economisation of the irrigation water. Another research by Janmohammadi et al. (2016b) in barley on chelated nano-ZnO showed significant increase in the days to anthesis and maturity after application, as well as enhancement in the yield-attributing characteristics and yield. Similarly, the foliar application of nano-zinc oxide in rice cv. PR-121 and Basmati-1509 enhanced the vegetative growth, yield and grain zinc content (Bala 2018).

6. How to Trace the Uptake Path of the Applied Nano-Nutrients?

The fate of the applied nano-nutrients has to be discerned in order to ensure its uptake as nano-crystalline entities by the plant cells. This can be done by use of advanced microscopy and spectroscopy tools and techniques which have been summarized in the following section.

6.1 Viewing Nano-Nutrient Adsorption and Uptake by Advanced Microscopy Tools

6.1.1 Advanced Fluorescence and Confocal Scanning Laser Microscopy (CSLM)

Tagging of fluorescent dyes (organic or natural ones) or nano-crystalline semi-conductor quantum dots exhibiting self-fluorescence to MN NPs can be beneficial in elucidating their movement from plant cell exterior to apoplast or symplast routes followed during adsorption, uptake and partitioning (Wu et al. 2017). The NP localization can be further improved with high resolution using time lapse CSLM study that involves point-to-point scanning of the specimen by a focused laser beam (Prasad et al. 2007). Similarly, the polymer nano-coated or nano-encapsulated micronutrients can be traced by adding organic fluorescent dyes, however, the size should not exceed the plant cell wall porosity dimensions (Albersheim et al. 2011). The nano-cell interface studies can be further improved by use of localized surface plasmon resonance (LSPR) on 2-D nobel metal NP sheets as the substrates in lieu of glass/quartz (Masuda et al. 2017).

6.1.2 High Resolution Transmission Electron Microscopy (HR-TEM) and Electron Beam Trapping Techniques

These are advanced techniques which enable the localization of structures within the cells with resolution upto atomic level. High resolution cryo-TEM can be used to obtain information about the agglomeration of the nanoparticles (Balmes et al. 2006). Another TEM variant involves trapping of NPs by focusing an electron beam on them, similar to optical tweezers (Balijepalli et al. 2012). Optical trapping can also be done with plasmonic nanostructures, such as silicon micro-ring resonators (Crozier 2016). This trapping ensures the localization of the dense, high electron contrast NPs in the biological systems (Balijepalli et al. 2012).

6.1.3 Aggregation Induced Emission (AIE)

This phenomenon is employed in order to overcome the aggregation caused quenching (ACQ) in plant cells when conventional fluorophores are used. Aggregation-induced emission occurs when intramolecular rotations of molecules (having rotating units) suffer restriction (Hong et al. 2009). The main mechanisms on which AIE functions are: Restriction of Intramolecular Rotations (RIR), Restriction of Intramolecular Vibrations (RIV) and Restriction of Intramolecular Motions (RIM) (Mei et al. 2015). AIE luminogens (AIEgens) provide a useful tool for chemical sensing in plant cells which involves ion detection. AIEgens are molecules with no emission property, owing to their free intramolecular motion, but they exhibit the phenomenon of fluorescence when they form an aggregate as a result of restriction to their motions (Qian and Tang 2017). Therefore, these can be used as bioprobes for their application in cell imaging.

6.2 Sensing Nano-Nutrients by Advanced Spectroscopy Techniques

6.2.1 Special Inductively Coupled Plasma-Mass Spectroscopy (ICP-MS)

The ICP-MS-based novel techniques can be employed for detection and quantification of metal elements at higher sensitivity and efficiency by using plasma generated by magnetic field of radiofrequency (Cubadda 2007). The Single Particle ICP-MS involves complete atomization of NPs in the plasma for generation of ion packets for detection in MS. The versatility of SP-ICP-MS is ultra-high detection sensitivity to identify amount of analyte in its solution form and as NP in complex matrices, such as plant cells,with cost-effective, easy sample preparation and high throughput analysis without requirement of any prior separation techniques (Bao et al. 2016). Moreover, it can also provide information on the size and size distribution of metal/metal oxide NPs. Another variant Laser Ablation (LP) ICP-MS is a step ahead of SP-ICP-MS to identify occurrence of element NP either in quantitative amounts or as spatially resolved mapping of elements in intricate plant or animal tissues (Pozebon et al. 2017).

6.2.2 Fourier Transform Infra-Red nano-Spectroscopy (FT-IRNS)

Nano-FTIR is advanced FT-IR exhibiting better spectral resolution (less than 20 nm) for locating the polymer or organic compound coated metal/metal oxide NPs in polymeric nanocomposites and other complex matrices, such as plant cells (Amenabar et al. 2017).

7. Possible Conjugate Micro-Spectroscopy Methods: The Chimeric Technologies

7.1 Time-Resolved Laser Induced Fluorescence Spectroscopy (TRLIFS)

This technique employs time-domain fluorescence microscopy in order to measure the intensity decay of the sample on UV illumination, i.e., the time period for fluorescence emission from sample (Butte and Mamelak 2012). Thus, it captures the emitted fluorescence lifetime, intensity, and spectra from an excitation which can be significantly affected by acquisition environments. The dual quantitative spectroscopy and qualitative microscopy modes can be helpful in localization and concentration or number of MN NPs on application in plant tissues.

7.2 Nano-SIMS

Nano secondary ion mass spectrometry delivers the elemental distribution of low detection limit particles by employing an ion beam to remove particles from sample surface (Wang et al. 2017) by a process known as sputtering. The ionized atoms or molecules of the sputtered material are then subjected to mass spectroscopy analysis. The incident ion beam and the transmitted secondary ion signals are fed into specific software in order to reconstruct the image of the sample. The beam employed in this technique has a much larger probe diameter (50 nm) as compared to that of an electron beam in scanning electron microscope and employs secondary ion signals for information rather than X-rays. The ion beam is composed of Cs^+ ion in the case of negative ion analysis, and O^- ions in the case of analysis of positive ions (Nunez et al. 2017). Many reports have been published regarding distribution of metal ions in plant cells studied via Nano-SIMS (Martin et al. 2008, Moore et al. 2010).

8. Skepticism, Knowledge Gaps and Future Prospects

The rising research interest, fund flow and patents filed or published on novel fertilizers, 'nanofertilizers', showcase their potentiality. However, a similar or even greater amount of published literature has shown the possibilities of negative or ill impacts on the rampant usage of these fertilizers. The unknown risks and toxicology of NPs need to be thoroughly evaluated before the commercialization of these MN nanofertilizers. Nanoparticles can enter human bodies through dermal uptake, the respiratory system, and through trophic transfer. Although certain MN nanofertilizers have shown promising effects on plant growth and yield components, their toxic nature at higher concentrations limit their use as fertilizers (Morales-Diaz et al. 2017, Mahawar et al. 2017, Peralta-Videa et al. 2014).

Commercialization of the newly-developed processes for nanofertilizer synthesis also impede the growth of the nanofertilizer industry. The transformation of NPs in soil needs to be extensively studied in order to synthesize and recommend better MN nanofertilizers for plant growth. Nevertheless, the opportunities for the nano-interventions in plant nutrition and fertilizers are growing. However, the dynamics of nano-nutrient element on application and the possible nano-toxicity issues have to be sincerely addressed.

9. Concluding Remarks

The nanofertilizers are the products of the nanotechnology interventions which are at the incipient stage of development. Scientific evidence and increased awareness in support of positive effects are out-weighing the apprehensions related to environmental hazards associated with the use of nano-scale nutrients at large scale. Thus, enhanced NUE, targeted placement and release of the loaded nutrients, plant health and yield improvement, and environmental security features of the nanofertilizers endear a new revolution in the fertilizer industry.

Acknowledgements

The primary author sincerely thanks ICAR, New Delhi, for providing the necessary funding under the Nanotechnology platform for the research work, and Head Department of Soil Science, PAU, Ludhiana, Punjab, India, for providing the necessary infrastructure facilities. The secondary author graciously thanks DST, New Delhi for providing the DST Inspire fellowship for pursuing of the doctoral research.

References

Ahanger, M.A., N. Morad-Talab, E.F. Abd-Allah, P. Ahmad and R. Hajiboland. 2016. Plant growth under drought stress: Significance of mineral nutrients. pp. 649–668. *In*: Ahmad, P. (ed.). Water Stress and Crop Plants: A Sustainable Approach. Volume 2, First Edition, John Wiley & Sons, Ltd.

Albersheim, P., A. Darvill, K. Roberts, R. Sederoff and A. Staehelin. 2011. Cell walls and plant anatomy. Plant Cell Walls, 241–242.

Amenabar, I., S. Poly, M. Goikoetxea, W. Nuansing, P. Lasch and R. Hillenbrand. 2017. Hyperspectral infrared nanoimaging of organic samples based on fourier transform infrared nanospectroscopy. Nature Comm. 8: 14402. doi:10.1038/ncomms14402.

Amorim, M., S. Gomes, A. Soares and J. Scott-Fordsmand. 2012. Energy basal levels and allocation among lipids, proteins, and carbohydrates in *Enchytraeus albidus*: Changes related to exposure to Cu salt and Cu nanoparticles. Water Air Soil Pollut. 223: 477–482.

Asadishad, B., S. Chahal, A. Akbari, V. Cianciarelli, M. Azodi, S. Ghoshal and N. Tufenkji. 2018. Amendment of agricultural soil with metal nanoparticles: Effects on soil enzyme activity and microbial community composition. Environ. Sci. Technol. doi:10.1021/acs.est.7b05389.

Ashfaq, M., N. Verma and S. Khan. 2016. Carbon nanofibers as a micronutrient carrier in plants: Efficient translocation and controlled release of Cu nanoparticles. Environmental Sci. Nano. doi:10.1039/c6en00385k.

Bai, C., C.C. Reilly and B.W. Wood. 2006. Nickel deficiency disrupts metabolism of ureides, amino acids, and organic acids of young pecan foliage. Plant Physiol. 140: 433–443.

Bala, R. 2018. Effect of zinc oxide micronutrient nanofertilizer on rice (*Oryza sativa*). M.S. Thesis, Punjab Agricultural University, Ludhiana, Punjab, India.

Balijepalli, A., J.J. Gorman, S.K. Gupta and T.W. LeBrun. 2012. Significantly improved trapping lifetime of nanoparticles in an optical trap using feedback control. Nano Lett. 12: 2347–2351. doi:dx.doi.org/10.1021/nl300301x.

Balmes, O., J.O. Malm, N. Pettersson, G. Karlsson and J.O. Bovin. 2006. Imaging atomic structure in metal nanoparticles using high-resolution cryo-TEM. Microsc. Microanal. 12: 145–150.

Bandyopadhyay, S., G. Plascencia-Villa, A. Mukherjee, C.M. Rico, M. Jose-Yacaman, J.R. Peralta-Videa and J.L. Gardea-Torresdey. 2015. Comparative phytotoxicity of ZnO NPs, bulk ZnO, and ionic zinc onto the alfalfa plants symbiotically associated with *Sinorhizobium mileloti* in soil. Sci. Total Environ. 515: 60–69.

Bao, D., Z.G. Oh and Z. Chen. 2016. Characterization of silver nanoparticles internalized by *Arabidopsis* plants using single particle ICP-MS analysis. Front. Plant Sci. 7: 32. doi:10.3389/fpls.2016.00032.

Behera, S.K. and R.K. Panda. 2009. Effect of fertilization and irrigation schedule on water and fertilizer solute transport for wheat crop in a sub-humid sub-tropical region. Agricul. Ecosystems Environ. 130(3): 141–155. doi:10.1016/j.agee.2008.12.009.

Ben-Moshe, T., S. Frenk, I. Dror, D. Minz and B. Berkowitz. 2013. Effects of metal oxide nanoparticles on soil properties. Chemosphere 90: 640–646.

Bindraban, P.S., C. Dimkpa, L. Nagarajan, A. Roy and R. Rabbinge. 2015. Revisiting fertilisers and fertilisation strategies for improved nutrient uptake by plants. Biol. Fertil. Soils. 51: 897–911. doi:10.1007/s00374-015-1039-7.

Blevins, D.G. and K.M. Lukaszewski. 1998. Boron in plant structure and function. Annu. Rev. Plant Physiol. Plant Mol. Biol., 481–500.

Boregowda, N., H. Puttaswamy, S.P. Harischandra, S.S. Hunthrike and G. Nagaraja. 2017. Trichogenic-selenium nanoparticles enhance disease suppressive ability of *Trichoderma* against downy mildew disease caused by *Sclerospora graminicola* in pearl millet. Scientific Rep. 7: 2612. doi:10.1038/s41598-017-02737-6.

Bouain, N., Z. Shahzad, A. Rouached, G.A. Khan, P. Berthomieu, C. Abdelly, Y. Poirier and H. Rouached. 2014. Phosphate and zinc transport and signalling in plants: Toward a better understanding of their homeostasis interaction. J. Exptl. Bot. 65(20): 5725–5741. doi:10.1093/jxb/eru314.

Burke, D.J., N. Pietrasiak, S.F. Situ, E.C. Abenojar, M. Porche, P. Kraj, Y. Lakliang and A.C.S. Samia. 2015. Iron oxide and titanium dioxide nanoparticle effects on plant performance and root associated microbes. Intl. J. Mol. Sci. 16: 23630–23650.

Burke, D.J., N. Pietrasiak, S.F. Situ, E.C. Abenojar, M. Porche, P. Kraj, Y. Lakliang and A.C.S. Samia. 2015. Iron oxide and titanium dioxide nanoparticle effects on plant performance and root associated microbes. Intl. J. Mol. Sci. 16: 23630–23650.

Burman, U., M. Saini and P. Kumar. 2013. Effect of zinc oxide nanoparticles on growth and antioxidant system of chickpea seedlings. Toxicol. Environ. Chem. 95: 605–616.

Butte, P. and A.N. Mamelak. 2012. Time-Resolved Laser Induced Fluorescence Spectroscopy (TRLIFS): A tool for intra-operative diagnosis of brain tumors and maximizing extent of surgical resection. pp. 161–172. *In*: Hayat, M.A. (ed.). Tumors of the Central Nervous System, Vol. 5, Springer, Dordrecht.

Cacique, I.S., G.P. Domiciano, F.A. Rodrigues and F.X.R. do Vale. 2012. Silicon and manganese on rice resistance to blast. Bragantia Campinas. 71(2): 239–244.

Calder, A.J., C.O. Dimkpa, J.E. McLean, D.W. Johnson and A.J. Anderson. 2012. Soil components mitigate the antimicrobial effects of silver nanoparticles towards a beneficial soil bacterium, *Pseudomonas chlororaphis* O6. Sci. Total Environ. 429: 215–222.

Carbone, S. 2014. Impact of metal and metal oxide engineered nanoparticles in soil and plant systems. Ph.D. Thesis, University of Bologna, Bologna.

Churchill, K.A. and H. Sze. 1984. Anion-sensitive, H-pumping ATPase of oat roots: Direct effects of Cl, NO(3), and a disulfonic stilbene. Plant Physiol. 76: 490–497.

Connorton, J.M., J. Balk and J. Rodríguez-Celma. 2017. Iron homeostasis in plants—A brief overview. Metallomics 9: 813–823. doi:10.1039/C7MT00136C.

Conway, J.R. and A.O. Keller. 2016. Gravity-driven transport of three engineered nanomaterials in unsaturated soils and their effects on soil pH and nutrient release. Water Res. 98: 250–260. doi:10.1016/j.watres.2016.04.021.

Cubadda, F. 2007. Inductively coupled plasma mass spectroscopy. pp. 697–751. *In*: Pico, Y. (ed.). Food Toxicants Analysis, Elsevier B.V.

Da Rocha Pinho, L.G., P.H. Monnerat, A.A. Pires, M.S.M. Freitas and C.R. Marciano. 2015. Diagnosis of boron deficiency in green dwarf coconut palm. Agric. Sci. 6: 164–174. doi:10.4236/as.2015.61015.

Dakora, F.D. and D.A. Phillips. 2002. Root exudates as mediators of mineral acquisition in low-nutrient environments. Plant Soil 245: 35–47.

Daraio, C. and S. Jin. 2012. Synthesis and patterning methods for nanostructures useful for biological applications. pp. 27–44. *In*: Silva, G.A. and V. Parpura (eds.). Nanotechnology for Biology and Medicine: At the Building Block Level, Fundamental Biomedical Technologies. Springer Science+Business Media, LLC. doi:10.1007/978-0-387-31296-5_2.

Davarpanah, S., A. Tehranifar, G. Davarynejad, J. Abadia and R. Khorasani. 2016. Effects of foliar applications of zinc and boron nano-fertilizers on pomegranate (*Punica granatum* cv. *Ardestani*) fruit yield and quality. Scientia Horticulturae 210: 1–8.

Dimkpa, C.O., D.E. Latta, J.E. McLean, D.W. Britt, M.I. Boyanov and A.J. Anderson. 2013. Fate of CuO and Zn Onano- and microparticles in the plant environment. Environ. Sci. Technol. 47(9): 4734–4742. doi:10.1021/es304736y.

Dimpka, C.O. and C.S. Bindraban. 2016. Fortification of micronutrients for efficient agronomic production: A review. Agron. Sustain. Dev. 36: 7. doi:10.1007/s13593-015-0346-6.

Disfani, M.N., A. Mikhak, M.Z. Kassaee and A. Maghari. 2016. Effects of nano Fe/SiO$_2$ fertilizers on germination and growth of barley and maize. Archs. Agron. Soil Sci. doi:10.1080/03650340.2016.1239016.

Dordas, C. 2008. Role of nutrients in controlling plant diseases in sustainable agriculture. A review. Agron. Sustain. Dev. 28: 33–46. doi:10.1051/agro:2007051.

Duhan, J.S., R. Kumar, N. Kumar, P. Kaur, K. Nehra and S. Duhan. 2017. Nanotechnology: The new perspective in precision agriculture. Biotechnol. Rep. 15: 11–23. doi:10.1016/j.btre.2017.03.002.

Elmer, W.H. and J.C. White. 2016. The use of metallic oxide nanoparticles to enhance growth of tomatoes and eggplants in disease infested soil or soilless medium. Environ. Sci. Nano. 3: 1072–1079.

Fageria, N.K., V.C. Baligar and R.B. Clark. 2002. Micronutrients in crop production. Adv. Agron. 77: 185–268.

Fawcett, D., J.J. Verduin, M. Shah, S.B. Sharma and G.E.J. Poinern. 2017. A review of current research into the biogenic synthesis of metal and metal oxide nanoparticles via marine algae and seagrasses. J. Nanosci. 2017: 8013850. https://doi.org/10.1155/2017/8013850.

Fitzpatrick, K.L., S.D. Tyerman and B.N. Kaiser. 2008. Molybdate transport through the plant sulfate transporter SHST1. FEBS Letters 582: 1508–1513.

Gasber, A., S. Klaumann, O. Trentmann, A. Trampczynska, S. Clemens, S. Schneider, N. Sauer, I. Feifer, F. Bittner and R.R. Mendel. 2011. Identification of an *Arabidopsis* solute carrier critical for intracellular transport and inter-organ allocation of molybdate. Plant Biol. 13(5): 710–718.

Ghosh, S., H. Mashayekhi, B. Pan, P. Bhowmik and B. Xing. 2008. Colloidal behavior of aluminium oxide nanoparticles as affected by pH and natural organic matter. Langmuir. 24: 12385–12391.

Gonzalez, L., C.M.J.R. Peralta-Videa and J.L. Gardea-Torresdey. 2010. Toxicity and biotransformation of uncoated and coated nickel hydroxide nanoparticles on mesquite plants. Environ. Toxicol. Chem. 29(5): 1146–1154.

Grusak, M.A. 2001. Plant macro- and micronutrient minerals. Encyc. Life Sci. pp. 1–6.

Gupta, U. and H. Solanki. 2013. Impact of boron deficiency on plant growth. Intl. J. Bioassays. 2(7): 1048–1050.

Hajiboland, R. 2012. Effect of micronutrient deficiencies on plants stress responses. pp. 283–329. *In*: Ahmad, P. and M.N.V. Prasad (eds.). Abiotic Stress Responses in Plants: Metabolism, Productivity and Sustainability. doi:10.1007/978-1-4614-0634-1_16.

Hansch, R. and R.R. Mendel. 2009. Physiological functions of mineral micronutrients. Curr. Opin. Plant Boil. 12: 259–266.

Haris, Z. and I. Ahmad. 2017. Impact of metal oxide nanoparticles on beneficial soil microorganisms and their secondary metabolites. Intl. J. Life Sci. Scienti. Res. 3(3): 1020–1030.

Hong, J., L. Wang, Y. Sun, L. Zhao, G. Niu, W. Tan, C.M. Ricod, J.R. Peralta-Videa and J.L. Gardea-Torresdey. 2015. Foliar applied nanoscale and microscale CeO^2 and CuO alter cucumber (*Cucumis sativus*) fruit quality. Sci. Total Environ. 563-564: 904–911.

Hong, Y., J.W.Y. Lam and B.Z. Tang. 2009. Aggregation-induced emission: phenomenon, mechanism and applications. Chem. Commun. 29: 4332–4353.

Hu, C.W., M. Li, Y.B. Cui, D.S. Li, J. Chen and L.Y. Yang. 2010. Toxicological effects of TiO$_2$ and ZnO nanoparticles in soil on earthworm *Eisenia foetida*. Soil Biol. Biochem. 42: 586–591.

Huffman, D.L. and T.V. O'Halloran. 2001. Function, structure, and mechanism of intracellular copper trafficking proteins. Annu. Rev. Biochem. 70: 677–701.

Janmohammadi, M., A. Navid, A.E. Segherloo and N. Sabaghnia. 2016a. Impact of nano-chelated micronutrients and biological fertilizers on growth performance and grain yield of maize under deficit irrigation condition. Biologija 62(2): 134–147.

Janmohammadi, M., T. Amanzadeh, N. Sabaghnia and S. Dashti. 2016b. Impact of foliar application of nano-micronutrient fertilizers and titanium dioxide nanoparticles on the growth and yield components of barley under supplemental irrigation. Acta Agriculturae Slovenica 107(2): 265–276.

Janmohammadi, M., N. Sabaghnia, A. Seifi and M. Pasandi. 2017. The impacts of nano-structured nutrients on chickpea performance under supplemental irrigation. Acta Universitatis Agriculturae Et Silviculturae Mendelianae Brunensis 65: 858–870.

Jeevanandam, J., Y.S. Chan and M.K. Danquah. 2016. Biosynthesis of metal and metal oxide nanoparticles. Chem. Bio. Eng. Rev. 3(2): 55–67.

Jiang, J., G. Oberdorster and P. Biswas. 2009. Characterization of size, surface charge, and agglomeration state of nanoparticle dispersions for toxicological studies. J. Nanoparticle Res. 11: 77–89.

Joy, E.J.M., A.J. Stein, S.D. Young, E.L. Ander, M.J. Watts and M.R. Broadley. 2015. Zinc-enriched fertilisers as a potential public health intervention in Africa. Plant Soil 389: 1–24. doi:10.1007/s11104-015-2430-8.

Kamran, S., I. Shahid, D.N. Baig, M. Rizwan, K.A. Malik and S. Mehnaz. 2017. Contribution of zinc solubilizing bacteria in growth promotion and zinc content of wheat. Front. Microbiol. 8: 2593. https://doi.org/10.3389/fmicb.2017.02593.

Karimi, E. and E.M. Fard. 2017. Nanomaterial effects on soil microorganisms. pp. 137–200. *In*: Ghorbanpour, M., K. Manika and A. Varma (eds.). Nanoscience and Plant-Soil Systems. Soil Biology, Vol. 48. Springer, Cham.

Kaur, M. and A. Kalia. 2016. Role of salt precursors for the synthesis of zinc oxide nanoparticles in imparting variable antimicrobial activity. J. Appl. Natural Sci. 8(2): 1039–1048.

Kaur, S., N. Kaur, K.H.M. Siddique and H. Nayyar. 2016. Beneficial elements for agricultural crops and their functional relevance in defence against stresses. Archs. Agron. Soil Sci. 62(7): 905–920. doi:https://doi.org/10.1080/03650340.2015.1101070.

Khare, P., M. Sonane, Y. Nagar, N. Moin, S. Ali, K.C. Gupta and A. Satish. 2015. Size dependent toxicity of zinc oxide nano-particles in soil nematode *Caenorhabditis elegans*. Nanotoxicol. 9: 423–432.

Krithika, S. and D. Balachandar. 2016. Expression of zinc transporter genes in rice as influenced by zinc-solubilizing *Enterobacter cloacae* strain ZSB14. Front. Plant Sci. 7:446. https://doi.org/10.3389/fpls.2016.00446.

Kusunoki, M. 2007. Mono-manganese mechanism of the photosystem II water splitting reaction by a unique Mn4Ca cluster. Biochim. Biophys. Acta. 1767: 484–492.

Lee, W.M., Y.J. An, H. Yoon and H.S. Kweon. 2008. Toxicity and bioavailability of copper nanoparticles to terrestrial plants *Phaseolus radiatus* (Mung bean) and *Triticum aestivum* (Wheat): Plant agar test for water-insoluble nanoparticles. Environ. Toxicol. Chem. 27: 1915–1921.

Li, B., M. Tester and M. Gilliham. 2017. Chloride on the move. Trends Plant Sci. 22(3): 236–248.

Liu, R., H. Zhang and R. Lal. 2016. Effects of stabilized nanoparticles of copper, zinc, manganese, and iron oxides in low concentrations on lettuce (*Lactuca sativa*) seed germination: Nanotoxicants or nanonutrients? Water Air Soil Pollut. 227: 42.

Liu, Y., E. Donner, E. Lombi, R.Y. Li, Z.C. Wu, F.J. Zhao and P. Wu. 2013. Assessing the contributions of lateral roots to element uptake in rice using an auxin-related lateral root mutant. Plant Soil 372: 125–136.

Mahajan, P., S.K. Dhoke and A.S. Khanna. 2011. Effect of nano-ZnO particle suspension on growth of mung (*Vigna radiata*) and gram (*Cicer arietinum*) seedlings using plant agar method. J. Nanotechnol. 2011: 696535. doi:10.1155/2011/ 696535.

Mahawar, H., R. Prasanna, K. Simranjit, S. Thapa, A. Kanchan, R. Singh, S.C. Kaushik, S. Singh and L. Nain. 2017. Deciphering the mode of interactions of nanoparticles with mung bean (*Vigna radiata* L.). Israel J. Plant Sci. https://doi.org/10.1080/07929978.2017.1288516.

Mahmoodi, P., M. Yarnia, R. Amirnia, A. Tarinejad and H. Mahmoodi. 2017. Comparison of the effect of nano urea and nano iron fertilizers with common chemical fertilizers on some growth traits and essential oil production of *Borago officinalis* L. J. Dairy Vet. Sci. 2(2): JDVS.MS.ID.555585 (2017).

Majumdar, S., I.C. Almeida, E.A. Arigi, H. Choi, N.C. VerBerkmoes, J. Trujillo-Reyes, J.P. Flores-Margez, J.C. White, J.R. Peralta-Videa and J.L. Gardea-Torresdey. 2015. Environmental effects of nanoceria on seed production of common bean (*Phaseolus vulgaris*): A proteomic analysis. Environ. Sci. Technol. 49(22): 13283–13293.

Makarov, V.V., A.J. Love, O.V. Sinitsyna, S.S. Makarova, I.V. Yaminsky, M.E. Taliansky and N.O. Kalinina. 2014. Green Nanotechnologies: Synthesis of metal nanoparticles using plants. Acta Naturae 6(20): 35–44.

Manjili, M.J., S. Bidarigh and E. Amiri. 2014. Study the effect of foliar application of nano-chelate molybdenum fertilizer on the yield and yield components of peanut. Biological Forum-An Intl. J. 6(2): 37–40.

Mankad, M., R.S. Fougat, A. Patel, P. Mankad, G. Patil and N. Subhash. 2017. Assessment of physiological and biochemical changes in rice seedlings exposed to bulk and nano iron particles. Intl. J. Pure Appl. Biosci. 5(4): 150–159.

Marschner, P. 2012. Marschner's mineral nutrition of higher plants, 3rd edn. Elsevier, Oxford.

Maruthupandy, M., Y. Zong, J.S. Chen, J.M. Song, H.L. Niu, C.J. Mao, S.Y. Zhang and Y.H. Shen. 2017. Synthesis of metal oxide nanoparticles (CuO and ZnO NPs) via biological template and their optical sensor applications. Appl. Surface Sci. 397: 167–174. doi:https://doi.org/10.1016/j.apsusc.2016.11.118.

Masuda, S., Y. Yanase, E. Usukura, S. Ryuzaki, P. Wang, K, Okamoto, T. Kuboki, S. Kidoaki and K. Tamada. 2017. High-resolution imaging of a cell-attached nanointerface using a gold-nanoparticle two-dimensional sheet. Sci. Rep. 7: 3720. doi:10.1038/s41598-017-04000.

Mei, J., N.L.C. Leung, R.T.K. Kwok, J.W.Y. Lam and B.Z. Tang. 2015. Aggregation-induced emission: Together we shine, united we soar! Chem. Rev. 115: 11718–11940.

Miri, A.H., E.S. Shakib, O. Ebrahimi and J. Sharifi-Rad. 2017. Impacts of nickel nanoparticles on growth characteristics, photosynthetic pigment content and antioxidant activity of *Coriandrum sativum* L. Orient J. Chem. 33(3): 1297–1303.

Miwa, K., M. Tanaka, T. Kamiya and T. Fujiwara. 2010. Molecular mechanisms of boron transport in plants: Involvement of *Arabidopsis* NIP5; 1 and NIP6; 1. Adv. Exp. Med. Biol. 679: 83–96.

Monreal, C.M., M. DeRosa, S.C. Mallubhotla, P.S. Bindraban and C. Dimkpa. 2016. Nanotechnologies for increasing the crop use efficiency of fertilizer-micronutrients. Biol. Fertil. Soils. 52: 423–437. doi:10.1007/s00374-015-1073-5.

Moore, K.L., M. Schroder, E. Lombi, F.J. Zhao, S.P. McGrath, M.J. Hawkesford, P.R. Shewry and C.R.M. Grovenor. 2010. Nano-SIMS analysis of arsenic and selenium in cereal grain. New Phytol. 185: 434–445.

Morales-Diaz, A.B., H. Ortega-Ortíz, A. Juárez-Maldonado, G. Cadenas-Pliego, S. González-Morales and A. Benavides-Mendoza. 2017. Application of nanoelements in plant nutrition and its impact in ecosystems. Adv. Nat. Sci: Nanosci. Nanotechnol. 8: 013001.

Nair, R., S.H. Varghese, B.G. Nair, T. Maekawa, Y. Yoshida and D.S. Kumar. 2010. Nanoparticulate material delivery to plants. Plant Sci. 179: 154–163.

Nie, Z., C. Hu, Q. Tan and X. Sun. 2016. Gene expression related to molybdenum enzyme biosynthesis in response to molybdenum deficiency in winter wheat. J. Soil Sci. Plant Nutr. 16(4): http://dx.doi.org/10.4067/S0718-95162016005000071.

Nishida, S., C. Tsuzuki, A. Kato, A. Aisu, J. Yoshida and T. Mizuno. 2011. AtIRT1, the primary iron uptake transporter in the root, mediates excess nickel accumulation in *Arabidopsis thaliana*. Plant Cell Physiol. 52(8): 1433–1442. doi:10.1093/pcp/pcr089.

Nunez, J., R. Renslow, J.B. Cliff and C.R. Anderton. 2017. NanoSIMS for biological applications: Current practices and analysis. Biointerphases 13(3): 03B301.

Oliver, M.A. and P.J. Gregory. 2015. Soil, food security and human health: A review. Euro. J. Soil Sci. 66: 257–276.

Peralta-Videa, J.R., J.A. Hernandez-Viezcas, L. Zhao, B.C. Diaz, Y. Ge, J.H. Priester, P.A. Holden and J.L. Gardea-Torresdey. 2014. Cerium dioxide and zinc oxide nanoparticles alter the nutritional value of soil cultivated soybean plants. Plant Physiol. Biochem. 80: 128–135.

Pozebon, D., G.L. Schefflera and V.L. Dressler. 2017. Recent applications of laser ablation inductively coupled plasma mass spectrometry (LA-ICP-MS) for biological sample analysis: A follow-up review. J. Anal. At. Spectrom. 32: 890–919.

Prasad, V., D. Semwogerere and E.R. Weeks. 2007. Confocal microscopy of colloids. J. Phys. Condens. Matter. 11: 113102.

Printz, B., S. Lutts, J.F. Hausman and K. Sergeant. 2016. Copper trafficking in plants and its implication on cell wall dynamics. Front. Plant Sci. 7: 601. doi:10.3389/fpls.2016.00601.

Qian, J. and B.Z. Tang. 2017. AIE luminogens for bioimaging and theranostics: From organelles to animals. Chem. 3: 56–91.

Rahimizadeh, M., D. Habibi, H. Madani, G.N. Mohammadi and A.M. Sabet. 2007. The effect of micronutrients on antioxidant enzymes metabolism in Sunflower (*Helianthus annaus* L.) under drought stress. Helia. 30(47): 167–174. doi:10.2298/HEL0747167R

Raliya, R., R. Nair, S. Chavalmane, W.N. Wang and P. Biswas. 2015. Mechanistic evaluation of translocation and physiological impact of titanium dioxide and zinc oxide nanoparticles on the tomato (*Solanum lycopersicum* L.) plant. Metallomics. 7: 1584–1594.

Rashid, M.I., T. Shahzad, I.M.I. Ismail, G.M. Shah and T. Almeelbi. 2017a. Zinc oxide nanoparticles affect carbon and nitrogen mineralization of *Phoenix dactylifera* leaf litter in a sandy soil. J. Hazard. Mater. 324(Pt B): 298–305.

Rashid, M.I., T. Shahzad, M. Shahid, M. Imran, J. Dhavamani, I.M.I. Ismail, J.M. Basahi and T. Almeelbi. 2017b. Toxicity of iron oxide nanoparticles to grass litter decomposition in a sandy soil. Scientific Repts. 7: 41965. doi:10.1038/srep41965.

Raven, K.P. and R.H. Loeppert. 1977. Trace element composition of fertilizers and soil amendments. J. Environ. Qual. 26: 551–557.

Rawat, S., V.L.R. Pullagurala, M. Hernandez-Molina, Y. Sun, G. Niu, J.A. Hernandez-Viezcas, J.R. Peralta-Videa and J.L. Gardea-Torresdey. 2018. Impacts of copper oxide nanoparticles on bell pepper (*Capsicum annum* L.) plants: A full life cycle study. Environ. Sci. Nano. 5: 83–95.

Refaai, M.M. 2014. Response of Zaghloul date palms grown under Minia region conditions to spraying wheat seed sprout extract and nano-boron. Stem Cell 5(4): 22–28.

Revati, A.K. and B.D. Pandey. 2011. Microbial synthesis of iron-based nanomaterials—A review. Bull. Mater. Sci. 34(2): 191–198.

Rico, C.M., S.C. Lee, R. Rubenecia, A. Mukherjee, J. Hong, J.R. Peralta-Videa and J.L. Gardea-Torresdey. 2014. Cerium oxide nanoparticles impact yield and modify nutritional parameters in wheat (*Triticum aestivum* L.). J. Agric. Food Chem. 62: 9669–9675. http://dx.doi.org/10.1021/jf503526r.

Rietra, R.P.J.J., M. Heinen, C.O. Dimkpa and P.S. Bindraban. 2015. Effects of nutrient antagonism and synergism on fertilizer use. VFRC Report 2015/5. Virtual Fertilizer Research Center, Washington, DC, USA, pp. 42.

Schwarz, G. and R.R. Mendel. 2006. Molybdenum cofactor biosynthesis and molybdenum enzymes. Annu. Rev. Plant Biol. 57: 623–647.

Sebastian, A., A. Nangia and M.N.V Prasad. 2017. Carbon-bound iron oxide nanoparticles prevent calcium induced iron deficiency in *Oryza sativa* L. J. Agric. Food Chem. 65(3): 557–564.

Shah, V. and I. Belozerova. 2009. Influence of metal nanoparticles on the soil microbial community and germination of lettuce seeds. Water Air Soil. Pollut. 197: 143–148.

Shahzad, Z. and A. Amtmann. 2017. Food for thought: How nutrients regulate root system architecture. Curr. Opinion Plant Biol. 39: 80–87.

Shao, J.F., N. Yamaji, R.F. Shen and J.F. Ma. 2017. The key to Mn homeostasis in plants: Regulation of Mn transporters. Trends Plant Sci. 22(3): 215–224. doi:10.1016/j.tplants.2016.12.005.

Shen, J., C. Li, G. Mi, L. Li, L. Yuan, R. Jiang and F. Zhang. 2013. Maximizing root/rhizosphere efficiency to improve crop productivity and nutrient use efficiency in intensive agriculture of China. J. Exptl. Bot. 64(5): 1181–1192. doi:10.1093/jxb/ers342.

Sinclair, A.H. and A.C. Edwards. 2008. Micronutrient deficiency problems in agricultural crops in Europe. pp. 225–244. *In*: Alloway, B.J. (ed.). Micronutrient Deficiencies in Global Crop Production. Springer, Dordrecht.

Slomberg, D.L. and M.H. Schoenfisch. 2012. Silica nanoparticle phytotoxicity to *Arabidopsis thaliana*. Environ. Sci. Technol. 46: 10247–10254.

Socha, A.L. and M.L. Guerinot. 2014. Mn-euvering manganese: The role of transporter gene family members in manganese uptake and mobilization in plants. Front Plant Sci. 5: 106. doi:10.3389/fpls.2014.00106.

Solanki, P., A. Bhargava, H. Chhipa, N. Jain and J. Panwar. 2015. Nano-fertilizers and their smart delivery systems. pp. 164–174. *In*: Rai, M., C. Ribeiro, L. Mattoso and N. Duran (eds.). Nanotechnologies in Food and Agriculture, Springer, Cham. https://doi.org/10.1007/978-3-319-14024-7_4.

Subbaiah, L.V., T.N. Prasad, T.G. Krishna, P. Sudhakar, B.R. Reddy and T. Pradeep. 2016. Novel effects of nanoparticulate delivery of zinc on growth, productivity, and zinc biofortification in maize (*Zea mays* L.). J. Agric. Food Chem. 64: 3778–3788.

Subramanian, K.S. and C.S. Rahale. 2013. Nano-fertilizers-synthesis, characterization and application. pp. 69–80. *In*: T. Adhikari and Subba Rao (eds.). Nanotechnology in Soil Science and Plant Nutrition. New India Publishing Agency, New Delhi, India.

Taran, N.Y., O.M. Gonchar, K.G. Lopatko, L.M. Batsmanova, M.V. Patyka and M.V. Volkogon. 2014. The effect of colloidal solution of molybdenum nanoparticles on the microbial composition in rhizosphere of *Cicer arietinum* L. Nanoscale Res. Lett. 9: 289.

Teske, S.S. and C.S. Detweiler. 2015. The biomechanisms of metal and metal-oxide nanoparticles' interactions with cells. Intl. J. Environ. Res. Public Health. 12: 1112–1134. doi:10.3390/ijerph120201112.

Tkacz, A. and P. Poole. 2015. Role of root microbiota in plant productivity. J. Exptl. Bot. 66(8): 2167–2175. doi:10.1093/jxb/erv157.

Tourinho, P.S., C.A.M. van Gestel, S. Lofts, C. Svendsen, A.M.V.M. Soares and S. Loureiro. 2012. Metal-based nanoparticles in soil: Fate, behavior, and effects on soil invertebrates. Env. Tox. Chem. 31(8): 1679–1692.

Uchida, R. 2000. Essential nutrients for plant growth: nutrient functions and deficiency symptoms. pp. 31–55. *In:* Silva, J.A. and R. Uchida (eds.). Plant Nutrient Management in Hawaii's Soils, Approaches for Tropical and Subtropical Agriculture. College of Tropical Agriculture and Human Resources, University of Hawaii at Manoa.

Veresogloua, S.D., E.K. Bartoab, G. Menexesc and M.C. Rillig. 2013. Fertilization affects severity of disease caused by fungal plant pathogens. Plant Pathol. 62: 961–969. doi:10.1111/ppa.12014.

Voortman, R. and P.S. Bindraban. 2015. Beyond N and P: Toward a land resource ecology perspective and impactful fertilizer interventions in Sub-Saharan Africa. VFRC Report 2015/1. Virtual Fertilizer Research Center, Washington, DC, USA, pp. 49.

Wang, H., R.L. Wick and B. Xing. 2009. Toxicity of nanoparticulate and bulk ZnO, Al_2O_3 and TiO_2 to the nematode *Caenorhabditis elegans*. Environ. Pollut. 157: 1171–1177.

Wang, J., X. Zhang, L. Li, K. Cheng, J. Zheng, J. Zheng, M. Shen, X. Liu and G. Pan. 2016b. Changes in micronutrient availability and plant uptake under simulated climate change in winter wheat field. J. Soils Sediments doi:10.1007/s11368-016-1464-8.

Wang, P., E. Lombi, F.J. Zhao and P.M. Kopittke. 2016. Nanotechnology: A new opportunity in plant sciences. Trends Plant Sci. 21(8): 699–712.

Wang, Y., K. Thorup-Kristensen, L.S. Jensen and J. Magid. 2016a. Vigorous root growth is a better indicator of early nutrient uptake than root hair traits in spring wheat grown under low fertility. Front. Plant Sci. 7: 865. doi:10.3389/fpls.2016.00865.

Waraich, E.A., R. Ahmad, A. Halim and T. Aziz. 2012. Alleviation of temperature stress by nutrient management in crop plants: A review. J. Soil Sci. Plant Nutr. 12(2): 221–244.

White, P.J. and M.R. Broadley. 2001. Chloride in soils and its uptake and movement within the plant: A review. Annals Bot. 88: 967–988.

Wu, H., I. Santana, J. Dansie and J.P. Giraldo. 2017. *In vivo* delivery of nanoparticles into plant leaves. Curr. Protocols Chem. Biol. 9: 269–284. doi:10.1002/cpch.29.

Xuan, W., T. Beeckman and G. Xu. 2017. Plant nitrogen nutrition: Sensing and signaling. Curr. Opin. Plant Biol. 39: 57–65.

Zielonka, A. and M. Klimek-Ochab. 2017. Fungal synthesis of size-defined nanoparticles. Adv. Nat. Sci: Nanosci. Nanotechnol. 8: 043001. https://doi.org/10.1088/2043-6254/aa84d4.

CHAPTER 6

Application of Metabolomics to Discover the Implications of Nanomaterials for Crop Plants

Yuxiong Huang,[1,2] *Lijuan Zhao*[1,2] and *Arturo A. Keller*[1,2,*]

1. Introduction

Significant progress has been made over the past decade in understanding many of the risks of many engineered nanomaterials (ENMs), including the development of predictive tools to model their release from various applications into the environment throughout their life-cycle (Keller and Lazareva 2014, Gottschalk and Nowack 2011, Nowack et al. 2012), their behavior once released into the environment (Garner and Keller 2014, Lowry and Casman 2009), and some of the major toxicological paradigms (Xia et al. 2013, Peralta-Videa et al. 2011, Sajid et al. 2015, Cristiana and Guerranti 2015, Khan et al. 2016, Skjolding et al. 2016, Simonin and Richaume 2015). The toxicological information has increased to the level where effects to a number of species can be used to develop species sensitivity distributions (Garner et al. 2015, Coll et al. 2016), which can serve to identify acceptable ecological risk levels. Nevertheless, there is a need to fill in important data gaps, such as determining release rates of nanomaterials from consumer products, transformation rates of the nanomaterials via dissolution, oxidation, sulfidation, natural coatings and other processes. Increasing use of ENMs in agriculture, paints and coatings and other applications that impact the environment directly makes it imperative to understand their impact. An area that requires even deeper understanding is

[1] Bren School of Environmental Science and Management, University of California at Santa Barbara, CA, USA 93106.
[2] University of California, Center for Environmental Implications of Nanotechnology, Santa Barbara, CA, USA 93106.
* Corresponding author: keller@bren.ucsb.edu

the potential beneficial or negative effect of exposure to released nanomaterials at the predicted low concentration levels (parts-per-billion, ppb and sub-ppb) for most environments. Some of these effects may be harnessed to produce desirable outcomes, for example, promoting a positive organismal response (higher biomass, faster growth). This requires novel approaches in order to elucidate more subtle responses than traditional toxicological endpoints.

In the past couple of decades, the use of advance techniques (the "omics") to study the response of organisms to stressors has developed very rapidly. Genomics, transcriptomics, proteomics and metabolomics now allow for a comprehensive understanding of how organisms interact with organic compounds (Bino et al. 2004, Nicholson et al. 1999, van Ravenzwaay et al. 2007, Ankley et al. 2006, Atha et al. 2012, Majumdar et al. 2015, Santos et al. 2010), such as growth substrates, vitamins, and signaling molecules, how these compounds interact within and between organisms and their immediate surroundings. Genomics, transcriptomics, and proteomics data provide information on how organisms may interact with these organic compounds. This information has led to key insights into biochemical pathways that are potentially active in the environment. The field of metabolomics complements these data because it is used to directly assess active biochemical pathways, by measuring the end products of biological metabolic activity, i.e., the metabolites.

While genomics, transcriptomics, and proteomics provide information on availability of genetic information to respond to stressors, the translation of the genes into mRNA, and the subsequent production of proteins, metabolomics provides key insights into the biochemical pathways that are influenced by exposure to a particular chemical. Metabolomics provides information on biological functions, but can also be used to assess health at the molecular level (Bundy et al. 2009). To date, most metabolomics studies have focused on low-molecular weight metabolites, although databases on all metabolites, including higher molecular weight organics, are growing very rapidly (Longnecker et al. 2015). Metabolites are the end product of gene expression (Fiehn 2002), and the changes of metabolites are regarded as ultimate responses of organisms to stress (Lu et al. 2013). Thus, environmental metabolomics is becoming a powerful tool for investigating the response of organisms to various stressors, e.g., water, light, temperature and high levels of metals (Obata and Fernie 2012). Metabolomics can be applied to a population (e.g., microbes), organisms, cells or even biofluids, making it a powerful approach to better understand the response at different levels (Allwood et al. 2007, Harrigan et al. 2007, Griffiths et al. 2010, Griffiths et al. 2007, Mashego et al. 2007, Aldridge and Rhee 2014).

Research on the metabolic response to exposure to nanomaterials and their transformation products has been mostly in medical applications and implications (Garcia-Contreras et al. 2015, Schnackenberg et al. 2012, Huang et al. 2012, Bu et al. 2010, Feng et al. 2013, Xu et al. 2013). Very few studies have been done on ecologically-relevant organisms (Aldridge and Rhee 2014, Li et al. 2015, MacCormack and Goss 2008). Moreover, much research has focused on effects at higher toxicities, resulting in an important gap in how engineered nanomaterials may affect organisms at lower exposures. However, there is great potential for applying metabolomics to better understand the effects of exposure to nanomaterials.

2. Metabolite Data Collection and Processing

Conceptually, implementing environmental metabolomics is relatively straight-forward. The organisms are exposed under controlled conditions to the chemical of interest, in this case the nanomaterials. Depending on the test organism, samples can be collected at different time points along the course of the exposure, which can provide a powerful insight into the onset of a response. Samples may include tissues (e.g., leaves, cells), organisms (e.g., microorganisms) or fluids and related material (e.g., root exudates). The samples are immediately freeze dried upon collection, in order to preserve the metabolites. The samples are then prepared for analysis in one of three platforms: ^{1}H (proton) NMR, GC-MS or LC-MS. NMR is the easiest in terms of sample preparation, although extraction of the metabolites requires careful selection of solvents (Li et al. 2015, Weljie et al. 2006, Viant et al. 2003, Lin et al. 2007, Samuelsson et al. 2006). However, NMR generally has low sensitivity and may not be useful for tracking small changes in metabolic profiles or for low exposure concentrations. GC-MS and LC-MS provide higher sensitivity, and can be applied in either untargeted (i.e., non-specific search for metabolites) or targeted (i.e., seeking previously identified chemicals) approaches (Dunn et al. 2013, Patti 2011, Dettmer et al. 2007, Junot et al. 2014, Lommen et al. 2007, Smart et al. 2010). However, many metabolites of interest need to be derivatized for use in GC-MS, which requires additional preparation steps, loss of mass and, thus, resolution (Wu et al. 2011, Gao et al. 2010, Xu et al. 2010, Kanani et al. 2008). Many metabolites are quite hydrophilic, making LC-MS an ideal platform for their analysis (Zhou et al. 2012, Theodoridis et al. 2008, Lu et al. 2008), in many cases without further processing after extracting the metabolites from the tissue using a suitable and compatible solvent.

3. Data Analysis

After metabolite identification and quantification, processing of the large dataset into meaningful results becomes the major task. The process has been considerably simplified due to the advent of a number of analytical tools that can serve to identify which metabolites have changed in a significant manner, what metabolic pathways are influenced and, thus, how the organism responds to the chemical stressor (Castillo et al. 2011, Brown et al. 2011). In particular, the development of MetaboAnalyst, a web-based metabolomic data processing tool, simplifies the massive data processing needs when hundreds (or thousands) of metabolites are identified (Xia et al. 2009). MetaboAnalyst can use different input data (NMR peak lists, binned spectra, MS peak lists, compound/concentration data). It can be used for data normalization, multivariate statistical analysis, graphing, metabolite identification and pathway mapping. The statistical tools can serve to run *t*-tests, Principal Component Analysis (PCA), Partial Least Squares Discriminant Analysis (PLS-DA), and hierarchical clustering, as demonstrated in Fig. 1.

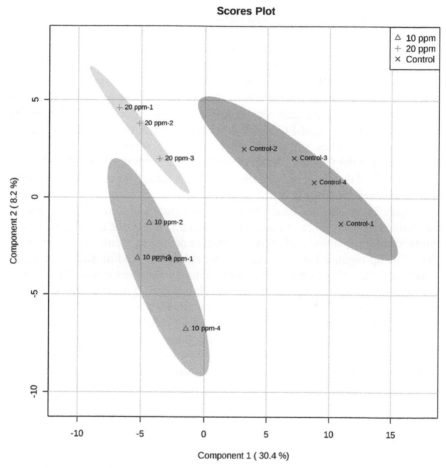

Fig. 1. Score plot of Principal Components 1 and 2 (PC1 vs PC2) using partial least squares-discriminant analysis (PLS-DA) analysis of metabolites in cucumber root exudates. Two-wk-old cucumber plants were exposed to different concentrations of nCu (0, 10, 20 mg/L) for one wk. Root exudates were collected for 4 hr at the end of nCu exposure and analyzed by GC-TOF-MS. PC1 explained 30.4% of the total variability. 56 metabolites were responsible for the separation. (With permission from (Zhou et al. 2016)).

4. Application of Metabolomics to Crop Plant Response to Nanopesticides

Our research group is at the forefront of the use of metabolomics to understand the effect of ENMs on organisms, particularly crop plants. We selected crop plants as our initial targets, in part because of the growing use of ENMs in agriculture, in part because engineered nanomaterials are directly applied to the plants, and in part because the applied concentrations are higher than in other ENM releases, providing a stronger signal of changes in metabolic profiles, for proof-of-concept. In prior studies, we exposed *Lactuca sativa* and *Cucumis sativus* to copper-based ENMs. The plants were grown in either natural soils or hydroponic media, and the exposure was either via foliar application (as recommended by the manufacturer) or via the soil (as would occur if biosolids were applied to the soil). We first evaluated the use of proton NMR vs. GC-MS based metabolomics (Zhao et al. 2016) in order to determine which platform would suit our work better. NMR identified 22 metabolites as significantly altered in the root exudates of cucumber plants exposed to nCu. Untargeted GC-TOF-MS analysis identified 156 metabolites in the same samples. Thus, while NMR provides some complementary information, it became clear the GC-MS would provide a richer dataset. Using PLS-DA, we then determined that there was clearly a significant difference between the control and the exposed groups (Fig. 1).

Figure 2 presents the most statistically significant up- and down-regulated metabolites after exposure of the plants to nCu and released Cu^{2+}. The increased exudation of some amino acids may be an active defense response of the cucumber plant. The up-regulated amino acids can provide many binding sites for copper, hindering the translocation from the root cell membrane. Previous studies showed that amino acids play an important role in chelating Cu^{2+} (Liao et al. 2000, Sharma and Dietz 2006). Based on the results of metabolomics, we hypothesized that the upregulation of these amino acids is an active defense of cucumber plants to excess nCu and Cu ions, and that by increasing these amino acids the plants would decrease Cu uptake. In order to verify this hypothesis, two-wk old cucumber plants were cultivated in 20 mg/L nCu nutrient solution with different levels of serine for 48 h. Interestingly, Cu accumulation in roots decreased with increasing serine concentration (Fig. 3). These results strongly indicate that amino acids are possibly released to detoxify nCu by binding with Cu ions (Zhao et al. 2016, Zhou et al. 2016). However, we cannot rule out the possibility that up-regulation of some amino acids was due to membrane damage, which caused leakage. These findings led us to consider whether other plants would have a similar behavior.

Several organic acids were also present at high concentrations in root exudates, particularly citric, succinic, malic and fumaric acids. However, only citric acid was significantly reduced after exposure to nCu, while changes in succinic, malic and fumaric acids were not statistically significant. The down-regulation of citric acid is possibly an active process to decrease the dissolution, uptake and translocation of Cu into cucumber tissues, or it could represent a shift in metabolism in the tricarboxylic acid cycle (TCA cycle). We hypothesized that citric acid plays a role in mobilizing Cu ions released from nCu, increasing Cu accumulation in cucumber plants. To confirm this hypothesis, we conducted an additional experiment in which we exposed cucumber plants to 20 mg/L nCu at different concentrations of citric acid. We found that the pH of the hydroponic system decreased from

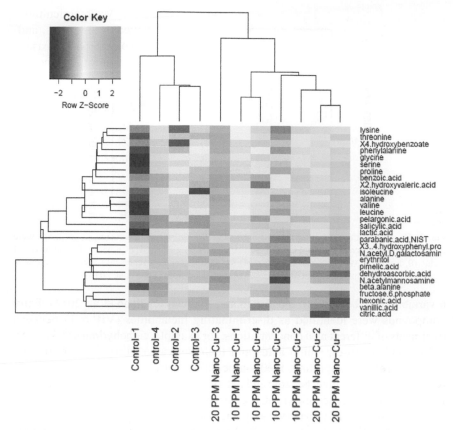

Fig. 2. Hierarchical cluster analysis of GC-MS data of cucumber root exudate. Centroid method and Euclidean distance were used for clustering. GC-MS data was log_2 transformed and raw scaled prior to cluster analysis. Red indicates a significant decrease in metabolite content, and green a significant increase in metabolite content. Dendrograms reveal relationships between different treatments based on metabolite abundance. Control samples were clearly separated from nCu treatments (With permission from (Zhou et al. 2016)).

6.37 (20 mg/L nCu without citric acid) to 5.28 (20 mg/L nCu with 6.25 mM citric acid). This decrease in pH led to increased dissolved Cu ions to concentrations that were 8 times higher than that in the control (20 mg/L nCu without citric acid). While further work is needed to verify these and other hypotheses resulting from the metabolomics analysis, this serves to illustrate the power of this approach in uncovering organismal responses to specific stressors.

In additional studies, with cucumber plants exposed throughout their entire life-cycle (seed to fruit harvest), we detected 239 metabolites using GC-TOF-MS, identifying 107 metabolites conclusively (An et al. 2016). The metabolites were ranked based on their variable importance in projection (VIP). VIP is the weighted sum of the squares of the PLS-DA analysis, which indicates the importance of

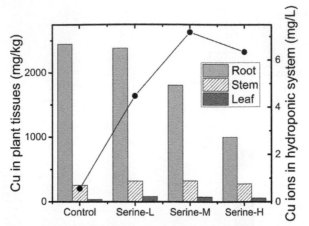

Fig. 3. Cu uptake in cucumber tissues and nutrient solution. Two-wk-old cucumber seedlings were cultivated in half strengh Hogland nutrient solution containing 20 mg/L nCu with different levels of serine (0, L = 6.25, M = 12.5, and H = 25 mM) for 48 h. Black dots represent Cu concentrations in nutrient solution at harvest. (With permission from (Zhao et al. 2016)).

a variable (metabolite) to the entire model (Garcia-Sevillano et al. 2014). Forty compounds were identified as discriminating metabolites (VIP > 1) between treatments of different nCu concentrations, for example, carbohydrates (1-kestose, xylose and fructose), amino acids and their derivatives (ornithine, citrulline, glycine, proline, oxoproline, methionine and aspartic acids), carboxylic acids (citric, glutaric, shikimic, benzoic and pelargonic acids) and fatty acids (arachidic, linolenic and caprylic acids). nCu clearly perturbed 15 metabolic pathways (Fig. 4). Five of these pathways (galactose metabolism, inositol phosphate metabolism, tricarboxylic acid (TCA) cycle, glyoxylate and dicarboxylate metabolism, starch and sucrose metabolism) are related to carbohydrate metabolism. Six pathways (arginine and proline metabolism; lysine biosynthesis; phenylalanine metabolism; phenylalanine, tyrosine and tryptophan biosynthesis; tyrosine metabolism; glycine, serine and threonine metabolism) are related to amino acid synthesis and metabolism. In addition, alpha-linolenic acid metabolism, isoquinoline alkaloid biosynthesis, and methane metabolism were also disturbed; these pathways are related to lipid metabolism, biosynthesis of other secondary metabolites, and energy metabolism, respectively. These results indicate that accumulated nCu may have a significant impact on carbohydrate and amino acid metabolism for cucumber plants. Interestingly, exposure to nCu reduced photosynthetic rate. In addition, stomatal conductance and transpiration rates tended to increase in nCu treatments compared to the control. The decline in carbon assimilation and increase in transpiration rates resulted in a statistically significantly decline in water use efficiency. However, root, stem, leaf and fruit biomass were not statistically impacted. Thus, these early results indicate the importance of understanding the extent to which organisms may be affected at lower doses than previously considered in previous studies of ENM ecotoxicity.

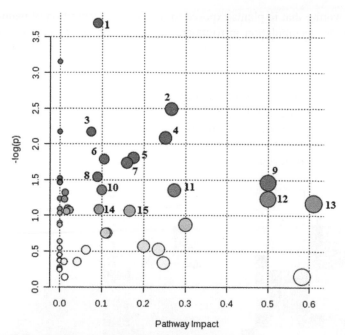

Fig. 4. Summary of pathway analysis with MetaboAnalyst 2.0. All identifed metabolites were considered in the pathway analysis. (1) Galactose metabolism; (2) Arginine & proline metabolism; (3) Lysine biosynthesis; (4) Inositol phosphate metabolism; (5) Citrate cycle (TCA cycle); (6) Glyoxylate & dicarboxylate metabolism; (7) alpha-Linolenic acid metabolism; (8) Starch & sucrose metabolism; (9) Phenylalanine metabolism; (10) Phenylalanine, tyrosine & tryptophan biosynthesis; (11) Tyrosine metabolism; (12) Isoquinoline alkaloid biosynthesis; (13) Glycine, serine & threonine metabolism; (14) Aminoacyl-tRNA biosynthesis; (15) Methane metabolism. (With permission from (An et al. 2016)).

A thorough and comprehensive analysis of metabolite changes of cucumber fruits from plants exposed to various levels of nano-Cu (0, 200, 400, and 800 mg kg^{-1}) was conducted through the use of both 1H NMR and GC–MS (Zhao et al. 2016). GC-MS detected and identified 107 metabolites in the cucumber fruit extract; 53 of them are related to nutritional supply, including sugars, amino acids, organic acids and fatty acids. Cucumber fruit extract contained 23 amino acids, among them, histidine, isoleucine, leucine, lysine, methionine, phenylalanine, threonine, tryptophan, tryptophan and valine, which are reported as essential human amino acids. Humans cannot synthesize these amino acids, therefore, they must be supplied via the diet (Shimomura et al. 2001). Among these essential amino acids, five amino acids (valine, leucine, isoleucine, threonine and tyrosine) were up-regulated in all nano-Cu treatments. Compared to the control, valine, leucine, isoleucine, threonine and tyrosine in fruits exposed to different concentrations of nano-Cu increased 16–32%, 13–41%, 12–28%, 0.1–18%, 0.1–32%, respectively.

It is noteworthy that in plants exposed to medium concentration of nano-Cu, the amino acids' accumulation in fruits are highest compared to control and other treatment. Among the essential amino acids, the content of lysine and methionine in fruit treated with different concentrations of nano-Cu decreased 55–61% and 13–25% respectively. However, all the changes are not statistically significant due to the large variation with groups. A correlation analysis (Fig. 5) showed that valine, leucine and isoleucine are clustered, indicating strong correlation. These three amino acids are commonly referred to as branched-chain amino acids (BCAAs) due to their branched carbon skeletons and play important roles in growth and development (Kochevenko et al. 2012).

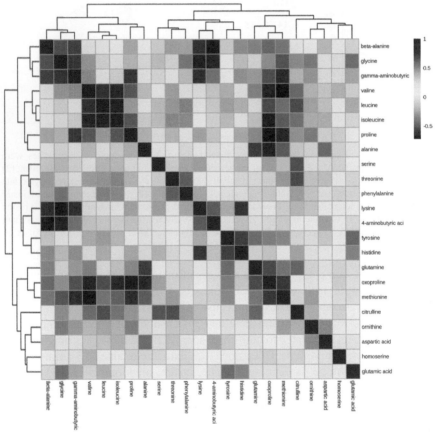

Fig. 5. Correlation map of amino acids in cucumber fruit extract. (With permission from (Zhao et al. 2016)).

In another study (Keller et al. 2016), lettuce plants were sprayed (foliar exposure) with Kocide, a $Cu(OH)_2$ nanomaterial. Using untargeted GC-TOF-MS, 352 compounds were detected and 159 metabolites were identified. PLS-DA analyses of all detected compounds were performed (Fig. 6), and 42 metabolites had VIP scores > 1, indicating these are the metabolites that play an important role in group separation. These 42 metabolites include carboxylic acids (fumaric, malic, maleic, oxalic, malonic, aconitic acids), amino acids (GABA, beta-alanine, glycine, tryptophan, proline, histidine, citrulline, alanine, aspartic acid, asparagine, oxoproline, serine, glutamic acid), amines (hydroxylamine, nicotianamine, ethanolamine), sugar (lyxose, 1-kestose), fatty acid (behenic acid) and other metabolites. Treated and control plants can be clearly separated via PLS-DA.

Pathway analysis indicated that six biological pathways were significantly perturbed in the lettuce leaves (Fig. 7A): (1) glycine, serine and threonine metabolism; (2) alanine, aspartate and glutamate metabolism; (3) tricarboxylic (TCA) cycle; (4) pantothenate and coenzyme-A (CoA) biosynthesis; (5) glycolysis or gluconeogenesis; and (6) pyruvate metabolism. $Cu(OH)_2$ NP treated plant leaves exhibited lower levels of TCA cycle intermediates, such as citric, isocitric and fumaric acids; downregulation of TCA cycle appears to be a clear response. Exposure to $Cu(OH)_2$ nanopesticide increased the levels of pyruvic acid 2–5 times compared to the control, indicating three biological pathways (4, 5, and 6 in Fig. 7) in which pyruvate participates were likely perturbed. In previous studies, soluble sugars were highly sensitive to environmental stress; these sugars play an important role in signaling and stress defense. In this study, sucrose, glucose, fructose and hexose concentrations did not change after exposure to $Cu(OH)_2$ nanopesticides.

Since exposure was through the leaves, Cu (probably in ionic form) was translocated to the roots (control experiments covering the soil to avoid deposition of $Cu(OH)_2$ nanopesticide were performed in order to ascertain that translocation did indeed occur). Although Cu concentration in the roots did increase relative to the unexposed controls (i.e., no foliar or soil application), the increase was moderate (from 6.0 ± 2.8 to 34.9 ± 10.4 mg/kg) compared to the increase in mesophyll tissue (from 13.0 ± 5.9 to $2,296 \pm 302$ mg/kg). Thus, metabolic profile changes were smaller, and affected different metabolic pathways than in the leaves (Fig. 6B), although in both cases the alanine, aspartate and glutamate metabolism pathway was altered. It should be noted that the threshold level for Cu ion to induce toxicity in some plants is 20–30 mg/kg (Marschner and Marschner 2012). However, high concentrations (823–2,501 mg/kg) of $Cu(OH)_2$ nanopesticide in the lettuce leaves did not cause any visible toxic symptoms throughout the entire exposure period. On the contrary, leaf biomass increased significantly. This suggested to us that there may be a beneficial dose level for certain ENMs.

In another set of experiments, 24-d-old lettuce plants grown in soil were exposed via the leaves to different concentrations (0, 1050 and 1555 mg/L) of $Cu(OH)_2$ nanopesticide for one month (Hannah-Bick et al. 2016). Results showed

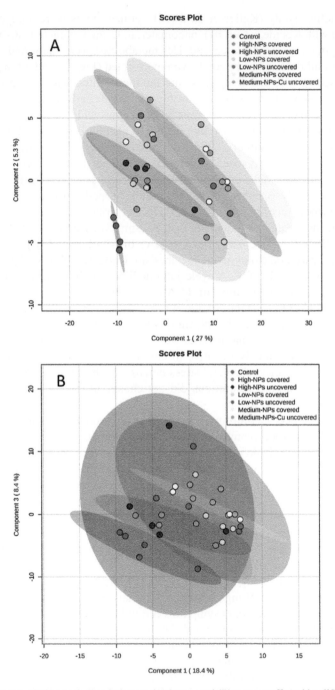

Fig. 6. PLS-DA of 159 metabolites in lettuce (A) leaves and (B) roots as affected by different doses of Cu(OH)$_2$ nanopesticide. (With permission from (Keller et al. 2016)).

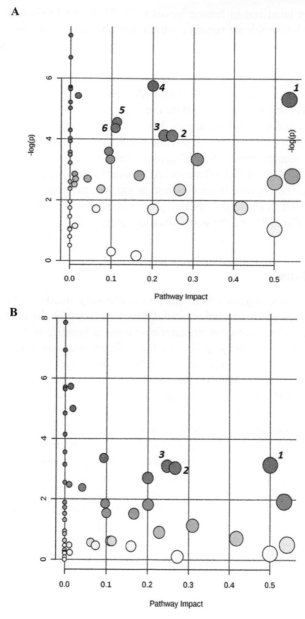

Fig. 7. Summary of pathway analysis with MetaboAnalyst 2.0 in lettuce (A) leaves and (B) roots. All detected metabolites were considered in the pathway analysis. Altered pathways in leaves: (1) glycine, serine and threonine metabolism; (2) alanine, aspartate and glutamate metabolism; (3) tricarboxylic (TCA) cycle; (4) pantothenate and coenzyme-A (CoA) biosynthesis; (5) glycolysis or gluconeogenesis; (6) pyruvate metabolism. Altered pathways in roots: (1) phenylalanine metabolism; (2) arginine and proline metabolism; (3) alanine, aspartate and glutamate metabolism. (With permission from (Keller et al. 2016)).

Cu was mainly localized in lettuce leaves (823–1111 and 1353–2008 mg/kg in vascular and photosynthetic tissues), which may potentially increase Cu intake and impact human health. In addition, foliar application of $Cu(OH)_2$ nanopesticide significantly increased potassium concentration in lettuce leaves by 6–7% and 21–28%, in vascular and photosynthetic tissues.

A total of 159 organic compounds were identified using GC-TOF-MS. A t-test was performed in order to identify metabolite concentrations that were significantly different between control and treatments. 39 compounds were significantly different compared to the control ($p < 0.01$). PCA analysis was shown in Fig. 8A, the control group and the nanopesticides treated groups were clearly separated along PC1, which explained 92% of the difference. Although the two treatment levels were separated from the control, there was not as much separation between low and high levels. PLS-DA showed that three groups were completely separated along component 1 (Fig. 8B) and the discriminating compounds were screened out by VIP value.

5. Conclusions

Our previous work suggests that there is a pressing need to conduct more systematic studies of the similarities and differences in metabolic profile changes between different engineered nanomaterials across a number of test organisms. Understanding how metabolic pathways for different organisms are affected by ENMs can have important implications in the fields of agriculture, ecology and environmental health science. As case studies, we evaluated the implications of copper-based ENMs (nano-Cu and $Cu(OH)_2$ nanoparticles) on two major crop plants, cucumber (*Cucumis sativus*) and lettuce (*lactuca sativa*) via non-targeted metabolomics to obtain a comprehensive understanding of the effects.

^{1}H NMR and GC–MS-based metabolomics, as well as ICP-MS-based metallomics, were used to demonstrate that exposure of plants to ENMs throughout their entire life cycle can result in significant changes in metabolite profiles and metal bioaccumulation. The subsequent metabolic pathway analysis revealed both carbon and nitrogen metabolisms were significantly disturbed, affecting the levels of amino acids, carbohydrates and other important biomolecules. Results also showed ENMs significantly altered some patterns in the metabolites, indicating that the nutrients supplied by the fruit will be changed. As ENMs may generate ROS, the plant appears to respond by using its antioxidant capacity, in the form of phenolic compounds, vitamins, and other metabolites, which results in a decrease in the total antioxidant capacity contained in the product to be consumed. The metabolite and metallomic profiling provide information that could be used to launch more detailed investigations of specific effects or mechanisms of response from exposure to nanomaterials and other chemicals. Furthermore, after having identified the metabolite changes induced by ENMs, one can study the up-stream genes controlling and regulating the identified metabolites, in order to provide

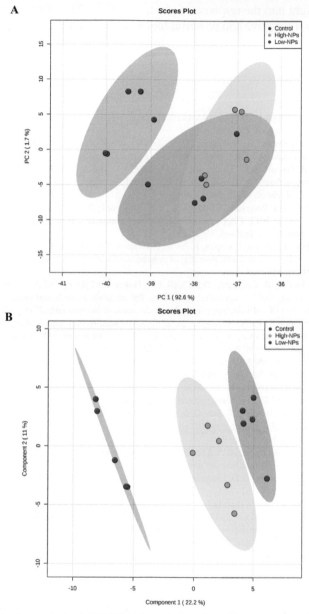

Fig. 8. Multivariate analysis of GC-TOF-MS metabolites data of lettuce photosynthesis tissues, A: Score plot of Principle Component Analysis (PCA); B: Score plot of Partial Least Square Discriminant Analysis (PLS-DA). After 24 d lettuce plants were exposed to different concentrations (0, 1050 and 1555 mg/L) of Cu(OH)$_2$ nanopesticide for one month. At harvest, lettuce leaves were collected and analyzed by GC-TOF-MS. Each treatment had 5 replicates. (with permission from (Hannah-Bick et al. 2016)).

additional insight into the response of plants to ENMs. Therefore, metabolomics can be used as a sensitive and powerful tool to understand the response of plants to nanoparticles at a molecular level, as an early indicator of potential implications from exposure to ENMs.

References

Aldridge, Bree B. and Kyu Y. Rhee. 2014. Microbial metabolomics: Innovation, application, insight. Curr. Opin. Microbiol. 19: 90–96.

Allwood, J.W., D.I. Ellis and R. Goodacre. 2007. Metabolomic technologies and their application to the study of plants and plant–host interactions. Physiol. Plant. 132.2: 117–135.

Ankley, G.T., G.P. Daston, S.J. Degitz, N.D. Denslow, R.A. Hoke, S.W. Kennedy, A.L. Miracle, E.J. Perkins, J. Snape, D.E. Tillitt et al. 2006. Toxicogenomics in regulatory ecotoxicology. Environ. Sci. Technol. 40(13): 4055–4065.

Atha, D.H., H. Wang, E.J. Petersen, D. Cleveland, R.D. Holbrook, P. Jaruga, M. Dizdaroglu, B. Xing and B.C. Nelson. 2012. Copper oxide nanoparticle mediated dna damage in terrestrial plant models. Environ. Sci. Technol. 46(3): 1819–1827.

Bino, R.J., R.D. Hall, O. Fiehn, J. Kopka, K. Saito, J. Draper, B.J. Nikolau, P. Mendes, U. Roessner-Tunali, M.H. Beale et al. 2004. Potential of metabolomics as a functional genomics tool. Trends Plant Sci. 9(9): 418–425.

Brown, M., D.C. Wedge, R. Goodacre, D.M. Kell, P.N. Baker, L.C. Kenny, M.A. Mamas, L. Neyses and W.B. Dunn. 2011. Automated workflows for accurate mass-based putative metabolite identification in LC/MS-derived metabolomic datasets. Bioinformatics 27(8): 1108–1112.

Bu, Q., G. Yan, P. Deng, F. Peng, H. Lin, Y. Xu, Z. Cao, T. Zhou, A. Xue and Y. Wang. 2010. NMR-based metabonomic study of the sub-acute toxicity of titanium dioxide nanoparticles in rats after oral administration. Nanotechnology 21(12): 125105.

Bundy, J.G., M.P. Davey and M.R. Viant. 2009. Environmental metabolomics: a critical review and future perspectives. Metabolomics 5(1): 3–21.

Castillo, S., P. Gopalacharyulu, L. Yetukuri and M. Orešič. 2011. Algorithms and tools for the preprocessing of LC–MS metabolomics data. Chemom. Intell. Lab. Syst. 108(1): 23–32.

Coll, C., D. Notter, F. Gottschalk, T. Sun, C. Som and B. Nowack. 2016. Probabilistic environmental risk assessment of five nanomaterials (nano-TiO $_2$, nano-Ag, nano-ZnO, CNT, and fullerenes). Nanotoxicology 10(4): 436–444.

Cristiana, M.R. and C. Guerranti. 2015. Ecotoxicity of nanoparticles in aquatic environments: a review based on multivariate statistics of meta-data. J. Environ. Anal. Chem. 2(4).

Dettmer, K., P.A. Aronov and B.D. Hammock. 2007. Mass spectrometry-based metabolomics. Mass Spectrom. Rev. 26(1): 51–78.

Dunn, W.B., A. Erban, R.J.M. Weber, D.J. Creek, M. Brown, R. Breitling, T. Hankemeier, R. Goodacre, S. Neumann, J. Kopka et al. 2013. Mass appeal: Metabolite identification in mass spectrometry-focused untargeted metabolomics. Metabolomics 9(S1): 44–66.

Feng, J., J. Li, H. Wu and Z. Chen. 2013. Metabolic responses of HeLa cells to silica nanoparticles by NMR-based metabolomic analyses. Metabolomics 9(4): 874–886.

Fiehn, O. 2002. Metabolomics—the link between genotypes and phenotypes. Plant Mol. Biol. 48(1/2): 155–171.

Gao, X., E. Pujos-Guillot and J.-L. Sébédio. 2010. Development of a quantitative metabolomic approach to study clinical human fecal water metabolome based on trimethylsilylation derivatization and GC/MS analysis. Anal. Chem. 82(15): 6447–6456.

Garcia-Contreras, R., M. Sugimoto, N. Umemura, M. Kaneko, Y. Hatakeyama, T. Soga, M. Tomita, R.J. Scougall-Vilchis, R. Contreras-Bulnes, H. Nakajima et al. 2015. Alteration of metabolomic profiles by titanium dioxide nanoparticles in human gingivitis model. Biomaterials 57: 33–40.

García-Sevillano, M.A., T. García-Barrera, F. Navarro, N. Abril, C. Pueyo, J. López-Barea and J.L. Gómez-Ariza. 2014. Use of metallomics and metabolomics to assess metal pollution in doñana national park (SW Spain). Environ. Sci. Technol. 48(14): 7747–7755.

Garner, K.L. and A.A. Keller. 2014. Emerging patterns for engineered nanomaterials in the environment: A review of fate and toxicity studies. J. Nanoparticle Res. 16(8): 2503.

Garner, K.L., S. Suh, H.S. Lenihan and A.A. Keller. 2015. Species sensitivity distributions for engineered nanomaterials. Environ. Sci. Technol. 49(9): 5753–5759.

Gottschalk, F. and B. Nowack. 2011. The release of engineered nanomaterials to the environment. J. Environ. Monit. 13: 1145–1155.

Griffiths, W.J., K. Karu, M. Hornshaw, G. Woffendin and Y. Wang. 2007. Metabolomics and metabolite profiling: Past heroes and future developments. Eur. J. Mass Spectrom. 13: 45–50.

Griffiths, W.J., T. Koal, Y. Wang, M. Kohl, D.P. Enot and H.-P. Deigner. 2010. Targeted metabolomics for biomarker discovery. Angew. Chemie Int. Ed. 49(32): 5426–5445.

Harrigan, G.G., S. Martino-Catt and K.C. Glenn. 2007. Metabolomics, metabolic diversity and genetic variation in crops. Metabolomics 3(3): 259–272.

Huang, S.-M., X. Zuo, J.J. Li, S.F.Y. Li, B.H. Bay and C.N. Ong. 2012. Metabolomics studies show dose-dependent toxicity induced by SiO_2 nanoparticles in MRC-5 human fetal lung fibroblasts. Adv. Healthc. Mater. 1(6): 779–784.

Junot, C., F. Fenaille, B. Colsch and F. Bécher. 2014. High resolution mass spectrometry based techniques at the crossroads of metabolic pathways. Mass Spectrom. Rev. 33(6): 471–500.

Kanani, H., P.K. Chrysanthopoulos and M.I. Klapa. 2008. Standardizing GC–MS metabolomics. J. Chromatogr. B 871(2): 191–201.

Keller, A.A. and A. Lazareva. 2014. Predicted releases of engineered nanomaterials: From global to regional to local. Environ. Sci. Tech. Lett. 1(1): 65–70.

Khan, N.S., A.K. Dixit and R. Mehta. 2016. Nanoparticle Toxicity in Water, Soil, Microbes, Plant and Animals, 277–309.

Kochevenko, A., W.L. Araújo, G.S. Maloney, D.M. Tieman, P.T. Do, M.G. Taylor, H.J. Klee and A.R. Fernie. 2012. Catabolism of branched chain amino acids supports respiration but not volatile synthesis in tomato fruits. Mol. Plant 5(2): 366–375.

Li, L., H. Wu, C. Ji, C.A.M. van Gestel, H.E. Allen and W.J.G.M. Peijnenburg. 2015. A metabolomic study on the responses of daphnia magna exposed to silver nitrate and coated silver nanoparticles. Ecotoxicol. Environ. Saf. 119: 66–73.

Liao, M.T., M.J. Hedley, D.J. Woolley, R.R. Brooks and M.A. Nichols. 2000. Copper uptake and translocation in chicory (*Cichorium intybus L.* cv Grasslands Puna) and tomato (*Lycopersicon esculentum Mill.* cv Rondy) plants grown in NFT system. II. The role of nicotianamine and histidine in xylem sap copper transport. Plant Soil 223(1/2): 245–254.

Lin, C.Y., H. Wu, R.S. Tjeerdema and M.R. Viant. 2007. Evaluation of metabolite extraction strategies from tissue samples using NMR metabolomics. Metabolomics 3(1): 55–67.

Lommen, A., G. van der Weg, M.C. van Engelen, G. Bor, L.A.P. Hoogenboom and M.W.F. Nielen. 2007. An untargeted metabolomics approach to contaminant analysis: Pinpointing potential unknown compounds. Anal. Chim. Acta 584(1): 43–49.

Longnecker, K., J. Futrelle, E. Coburn, M.C. Kido Soule and E.B. Kujawinski. 2015. Environmental metabolomics: Databases and tools for data analysis. Mar. Chem. 177: 366–373.

Lowry, G.V. and E.A. Casman. 2009. Nanomaterial transport, transformation, and fate in the environment: A risk-based perspective on research needs. pp. 125–137. *In*: Igor Linkov and Jeffery Steevens (eds.). Nanomaterials: Risks and Benefits; Springer Netherlands: The Netherlands. Https://link. springer.com/book/10.1007/978-1-4020-9491-0.

Lu, W., B.D. Bennett and J.D. Rabinowitz. 2008. Analytical strategies for LC–MS-based targeted metabolomics. J. Chromatogr. B 871(2): 236–242.

Lu, Y., H. Lam, E. Pi, Q. Zhan, S. Tsai, C. Wang, Y. Kwan and S. Ngai. 2013. Comparative metabolomics in *Glycine max* and *Glycine soja* under salt stress to reveal the phenotypes of their offspring. J. Agric. Food Chem. 61(36): 8711–8721.

MacCormack, T.J. and G.G. Goss. 2008. Identifying and predicting biological risks associated with manufactured nanoparticles in aquatic ecosystems. J. Ind. Ecol. 12(3): 286–296.

Majumdar, S., I.C. Almeida, E.A. Arigi, H. Choi, N.C. VerBerkmoes, J. Trujillo-Reyes, J.P. Flores-Margez, J.C. White, J.R. Peralta-Videa and J.L. Gardea-Torresdey. 2015. Environmental Effects of nanoceria on seed production of common bean (*Phaseolus vulgaris*): A proteomic analysis. Environ. Sci. Technol. 49(22): 13283–13293.

Marschner, H. and P. Marschner. 2012. Mineral nutrition of higher plants; Academic Press.

Mashego, M.R., K. Rumbold, M. De Mey, E. Vandamme, W. Soetaert and J.J. Heijnen. 2007. Microbial metabolomics: Past, present and future methodologies. Biotechnol. Lett. 29(1): 1–16.

Nicholson, J.K., J.C. Lindon and E. Holmes. 1999. "Metabonomics": Understanding the metabolic responses of living systems to pathophysiological stimuli via multivariate statistical analysis of biological NMR spectroscopic data. Xenobiotica 29(11): 1181–1189.

Nowack, B., J.F. Ranville, S. Diamond, J.A. Gallego-Urrea, C. Metcalfe, J. Rose, N. Horne, A.A. Koelmans and S.J. Klaine. 2012. Potential scenarios for nanomaterial release and subsequent alteration in the environment. Environ. Toxicol. Chem. 31: 50–59.

Obata, T. and A.R. Fernie. 2012. The use of metabolomics to dissect plant responses to abiotic stresses. Cell. Mol. Life Sci. 69(19): 3225–3243.

Patti, G.J. 2011. Separation strategies for untargeted metabolomics. J. Sep. Sci. 34(24): 3460–3469.

Peralta-Videa, J.R., L. Zhao, M.L. Lopez-Moreno, G. de la Rosa, J. Hong and J.L. Gardea-Torresdey. 2011. Nanomaterials and the environment: A review for the biennium 2008–2010. J. Hazard. Mater. 186: 1–15.

Sajid, M., M. Ilyas, C. Basheer, M. Tariq, M. Daud, N. Baig and F. Shehzad. 2015. Impact of nanoparticles on human and environment: Review of toxicity factors, exposures, control strategies, and future prospects. Environ. Sci. Pollut. Res. 22(6): 4122–4143.

Samuelsson, L.M., L. Förlin, G. Karlsson, M. Adolfsson-Erici and D.G.J. Larsson. 2006. Using NMR metabolomics to identify responses of an environmental estrogen in blood plasma of fish. Aquat. Toxicol. 78(4): 341–349.

Santos, E.M., J.S. Ball, T.D. Williams, H. Wu, F. Ortega, R. van Aerle, I. Katsiadaki, F. Falciani, M.R. Viant, J.K. Chipman et al. 2010. Identifying health impacts of exposure to copper using transcriptomics and metabolomics in a fish model. Environ. Sci. Technol. 44(2): 820–826.

Schnackenberg, L.K., J. Sun and R.D. Beger. 2012. Metabolomics Techniques in Nanotoxicology Studies, pp. 141–156.

Sharma, S.S. and K.-J. Dietz. 2006. The significance of amino acids and amino acid-derived molecules in plant responses and adaptation to heavy metal stress. J. Exp. Bot. 57(4): 711–726.

Shimomura, Y., M. Obayashi, T. Murakami and R.A. Harris. 2001. Regulation of branched-chain amino acid catabolism: Nutritional and hormonal regulation of activity and expression of the branched-chain α-keto acid dehydrogenase kinase. Curr. Opin. Clin. Nutr. Metab. Care 4(5): 419–423.

Simonin, M. and A. Richaume. 2015. Impact of engineered nanoparticles on the activity, abundance, and diversity of soil microbial communities: A review. Environ. Sci. Pollut. Res. 22(18): 13710–13723.

Skjolding, L.M., S.N. Sørensen, N.B. Hartmann, R. Hjorth, S.F. Hansen and A. Baun. 2016. A critical review of aquatic ecotoxicity testing of nanoparticles—the quest for disclosing nanoparticle effects. Angew. Chemie.

Smart, K.F., R.B.M. Aggio, J.R. Van Houtte and S.G. Villas-Bôas. 2010. Analytical platform for metabolome analysis of microbial cells using methyl chloroformate derivatization followed by gas chromatography–mass spectrometry. Nat. Protoc. 5(10): 1709–1729.

Theodoridis, G., H.G. Gika and I.D. Wilson. 2008. LC-MS-based methodology for global metabolite profiling in metabonomics/metabolomics. TrAC Trends Anal. Chem. 27(3): 251–260.

van Ravenzwaay, B., G.C.-P. Cunha, E. Leibold, R. Looser, W. Mellert, A. Prokoudine, T. Walk and J. Wiemer. 2007. The use of metabolomics for the discovery of new biomarkers of effect. Toxicol. Lett. 172(1): 21–28.

Viant, M.R., E.S. Rosenblum and R.S. Tjeerdema. 2003. NMR-based metabolomics: a powerful approach for characterizing the effects of environmental stressors on organism health. Environ. Sci. Technol. 37(21): 4982–4989.

Weljie, A.M., J. Newton, P. Mercier, E. Carlson and C.M. Slupsky. 2006. Targeted profiling: Quantitative analysis of 1H NMR metabolomics data. Anal. Chem. 78(13): 4430–4442.

Wu, H., T. Liu, C. Ma, R. Xue, C. Deng, H. Zeng and X. Shen. 2011. GC/MS-based metabolomic approach to validate the role of urinary sarcosine and target biomarkers for human prostate cancer by microwave-assisted derivatization. Anal. Bioanal. Chem. 401(2): 635–646.

Xia, J., N. Psychogios, N. Young and D.S. Wishart. 2009. MetaboAnalyst: A web server for metabolomic data analysis and interpretation. Nucleic Acids Res. 37(Web Server), W652–W660.

Xia, T., D. Malasarn, S. Lin, Z. Ji, H. Zhang, R.J. Miller, A.A. Keller, R.M. Nisbet, B.H. Harthorn, H.A. Godwin et al. 2013. Implementation of a multidisciplinary approach to solve complex nano EHS problems by the UC center for the environmental implications of nanotechnology. Small 9(9-10): 1428–1443.

Xu, F., L. Zou, C.N. Ong, L. Zou, C.N. Ong and C.N. Ong. 2010. Experiment-originated variations, and multi-peak and multi-origination phenomena in derivatization-based GC-MS metabolomics. TrAC Trends Anal. Chem. 29(3): 269–280.

Xu, J., Y. Chen, R. Zhang, Y. Song, J. Cao, N. Bi, J. Wang, J. He, J. Bai, L. Dong et al. 2013. Global and targeted metabolomics of esophageal squamous cell carcinoma discovers potential diagnostic and therapeutic biomarkers. Mol. Cell. Proteomics 12(5): 1306–1318.

Zhao, L., C. Ortiz, A.S. Adeleye, Q. Hu, H. Zhou, Y. Huang and A.A. Keller. 2016. Metabolomics to detect response of lettuce (*Lactuca sativa*) to $Cu(OH)_2$ nanopesticides: Oxidative stress response and detoxification mechanisms. Environ. Sci. Technol. 50(17): 9697–9707.

Zhao, L., J. Hu, Y. Huang, H. Wang, A. Adeleye, C. Ortiz and A.A. Keller. 2006. 1H NMR and GC-MS based metabolomics reveal nano-Cu altered cucumber (*Cucumis sativus*) fruit nutritional supply. Plant Physiol. Biochem.

Zhao, L., Y. Huang, C. Hannah-Bick, A.N. Fulton and A.A. Keller. 2016. Application of metabolomics to assess the impact of $Cu(OH)_2$ nanopesticide on the nutritional value of lettuce (*Lactuca sativa*): Enhanced Cu intake and reduced antioxidants. NanoImpact 3: 58–66.

Zhao, L., Y. Huang, J. Hu, H. Zhou, A.S. Adeleye and A.A. Keller. 2016. 1H NMR and GC-MS based metabolomics reveal defense and detoxification mechanism of cucumber plant under Nano-Cu stress. Environ. Sci. Technol. 50(4): 2000–2010.

Zhao, L.J., Y. Huang, H. Zhou, A.S. Adeleye, H. Wang, C. Ortiz, S.J. Mazer and A.A. Keller. 2006. GC-TOF-MS based metabolomics and ICP-MS based metallomics of cucumber (*Cucumis sativus*) fruits reveal alteration of metabolites profile and biological pathway disruption induced by nano copper. Environ. Sci. Nano No. 5.

Zhou, B., J.F. Xiao, L. Tuli and H.W. Ressom. 2012. LC-MS-based metabolomics. Mol. BioSyst. 8(2): 470–481.

CHAPTER 7

Chitosan Nanomaterials for Smart Delivery of Bioactive Compounds in Agriculture

Ram Chandra Choudhary,[1] Sarita Kumari,[1]
R.V. Kumaraswamy,[1] Garima Sharma,[1] Ashok Kumar,[1,2]
Savita Budhwar,[2] Ajay Pal,[3] Ramesh Raliya,[4] Pratim Biswas[4]
and *Vinod Saharan[1,]**

1. Introduction

Global crop production is being threatened by climate change and one of the most important challenges is to supply a sufficient amount of food to the ever-increasing world population. Greenhouse gases are accumulating in the environment at unexpected rates and will increase the Earth's temperature by 2.5 to 4.3°C by 2080 in the crop growing areas of the world. Rising temperatures, decreasing water resources, floods, desertification and extreme weather will severely affect crop production and cause food shortages (Wang et al. 2014).

In agriculture, agrochemicals provide the primary means for better plant growth, control plant diseases and result in higher grain yield. However, continuous use of such chemicals to gain higher crop production causes several adverse effects (Carvalho 2006). Worldwide, ~ 2.5 million tons of pesticides are used each

[1] Department of Molecular Biology and Biotechnology, Rajasthan College of Agriculture, Maharana Pratap University of Agriculture and Technology, Udaipur, Rajasthan 313001, India.
[2] School of Interdisciplinary and Applied Life Sciences, Central University of Haryana, Mahendergarh, Haryana 123031, India.
[3] Department of Biochemistry, College of Basic Sciences and Humanities, Chaudhary Charan Singh Haryana Agricultural University, Hisar, Haryana 125004, India.
[4] Department of Energy, Environmental and Chemical Engineering, Washington University in St. Louis, MO 63130, USA.
* Corresponding author: vinodsaharan@gmail.com

year (De et al. 2014). More uses of these chemicals increase pest resistance and environmental pollution, decrease food quality and cause long-term implications for non-targeted organisms (Kohler and Triebskorn 2013). It has been estimated that more than 90 percent of applied agrochemicals are being lost as run-off or to the air during application,this severely impacts sustainability of agriculture and increases application cost to the farmers (Campos et al. 2015). Reducing the loss of agrochemicals has the potential to augment cost savings and increase sustainability in agriculture. Hence, there is growing emphasis on environmentally friendly technologies in agriculture and evaluation of various alternatives to reduce dependency on harmful agrochemicals.

The rapid progress in nanotechnology research has sought considerable attention in agricultural and food science through converging science and engineering. In agriculture, nanomaterial shave potential applications for solving problems by controlled release of encapsulated fertilizers, micronutrients, pesticides and detection of plant diseases, pollutants, pests and pathogens (Ghormade et al. 2011). The potential application of nanomaterials in plant growth and protection helps in the development of efficient and potential approaches for higher crop growth and control of plant pathogens. The application of nanomaterials in plants is largely associated with alteration of gene expression and biological pathways which ultimately affect plant growth and development (Nair et al. 2010). Nanomaterials can have varied compositions,such as metal oxide, silicates, ceramics, magnetic materials, dendrimers, polymers, liposome and emulsions. The composition of nanoparticles shows important starring role in their usage,for example,encapsulated polymeric nanoparticles are used as carrier for agrochemicals due to their controlled and slow release ability, whereas metal-based nanoparticles show size-dependent properties (Ghormade et al. 2011).

In recent years, nano-based smart delivery systems are most encouraging in the field of agriculture. Bio-degradable nanoparticles are gaining more attention due to site specific delivery of various biologically active compounds, such as plant growth regulators, vitamins and macro- and micronutrients (Tarafdar et al. 2013). Several bio-degradable and ecofriendly biomaterials of polysaccharide, lipid and protein nature are available as carriers for controlled delivery of bioactive agrochemicals. Among them, chitosan-based nanomaterials, such as nanoparticles, nanogels and nanocomposites, have attracted considerable attention in agriculture due to their antimicrobial, plant growth promoting, immune boosting and disease controlling properties (Kumaraswamy et al. 2018). However, selection of core material depends on various factors, such as encapsulating material, surface properties, size and biocompatibility.

2. Smart Delivery System for Agriculture

Nanotechnology-based delivery of biologically active ingredients to target site by carrier is one of the key methods in agriculture being used to address the current challenges caused by adverse environmental conditions and increased demand of food supply. The potential of nanomaterials has got increased consideration to improve current agriculture practices by enhancement of crop production, disease management, fisheries and animal production (Thornton 2010). Application of agrochemicals faces problems of over-application and reduced bioavailability due to soil chelation and run-offs. Therefore, there is a need to focus on the efficacy enhancement through controlled delivery systems for such agrochemicals. In recent years, the application of nanomaterials are increasing in crop improvement through delivery of pesticide/bio-pesticide encapsulated in nanomaterials for controlled release, slow release of fertilizer/bio-fertilizer and micronutrient encapsulation for efficient use, stabilization of plant growth regulators and targeted delivery of genetic material (Kashyap et al. 2015). Controlled release of biologically active ingredients and detection of pathogens by use of nanosensors are some of the progresses in food sector (Singh et al. 2017).

Smart delivery systems aim to achieve adequate and controlled release of required agrochemicals over a period of time and higher biological use fulness along with minimum harmful effects (Tsuji 2001). For this purpose, the use of several types of micro and nanoparticles as agrochemical delivery vehicles have been explored. As compared to micro size particles, nano size particles offer several advantages, such as high surface area, higher activity and fast mass transfer. Encapsulation of bioactive compounds within nanomaterials can be done through coating with thin polymer film, encapsulation inside nano-porous materials and development of nano-emulsion (Rai et al. 2012). Encapsulation plays an important role in protecting the environment by reducing evaporation and leaching through slow release mechanisms. Various types of nano-delivery systems, including emulsion, hydrogel, vesicular (liposome, noisome and transferosomes) and polymeric-based (chitosan, chitin, polyhydroxybutyrate, cellulose and starch) have been developed for agricultural applications (Parisi et al. 2015) (Fig. 1).

2.1 Nanoemulsion-Based Delivery System

Nanoemulsions are nanoscale droplets (oil/water system) of size less than ~ 100 nm (Anton and Vandamme 2011). Nanoemulsions have novel applications in various fields, such as agrochemicals, foods, cosmetics and pharmaceuticals. A typical nanoemulsion contains oil, water and emulsifier. They act as efficient delivery systems for hydrophobic compounds by dispersing the lipid phase as a colloidal dispersion. Nanoemulsions are commercially valuable delivery systems because they have the unique characteristic of small size and high surface area, transparent appearance, optical clarity and reduced rate of gravitational separation and flocculation. These are commonly stabilized by amphiphilic surfactants or emulsifiers which get absorbed between water and oil phases (Fig. 2). The emulsifier plays an important role for the creation of small sized droplets (Kumari et al. 2018). It reduces interfacial tension between oil and water phase and decreases the rate of coalescence of oil droplets by forming a physical, steric and/or electric barrier around them. Generally, synthetic surfactants, such as Tween-20 and Tween-80,

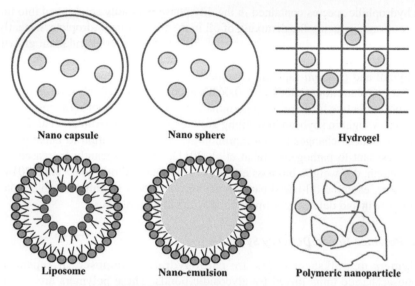

Fig. 1. Commonly-used nano-delivery systems for different purposes.

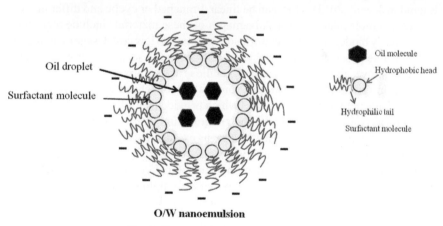

Fig. 2. Structure of oil/water nanoemulsion.

are used in emulsion. However, their usage increases the synthetic content in the emulsions, making them unsuitable for agriculture and the food industry, and also bearing environmental consequences.

Methods for the preparation of nanoemulsions are mainly divided into two categories: High energy approaches and low energy approaches. The high energy approaches disrupt the oil and aqueous phases into tiny droplets using mechanical devices such as high-pressure homogenizers, micro fluidizers and ultrasonicators (Mason 2006). In low energy methods, emulsion is formed spontaneously by mixing all the ingredients (oil, water, surfactant and/or co-surfactant) together. Further, the hydrophilic species contained in the oily phase is rapidly solubilized into the aqueous one, inducing the demixing of oil in the form of nano-droplets, instantly stabilized by the amphiphiles (Gupta et al. 2016). Antibacterial and plant growth promoting activities of nanoemulsion prepared using plant-based components, such as thymol and *Quillaja* saponin, have been tested on soybean crop. Among the various concentrations tested (0.01–0.06%), nanoemulsion at 0.03% concentration was most effective for improving germination percentage, shoot-root length, root and fresh weight in soybean in seedling growth experiments (Kumari et al. 2018). Thymol induced changes in root morphology, with the intention to make plants more resistant to pathogens (Tahat et al. 2011), and significantly increased the stem length, dry weight, biomass and yield of tomato (Abo-Elyousr et al. 2014). Therefore, essential oil-based nanoemulsions could be a potential antimicrobial, plant growth and defense system promoting agent in agriculture.

2.2 Polymer-Based Delivery System

Polymers or polysaccharides are carbohydrates composed of repeating monosaccharide units linked by glycosidic bonds. These polymers are widely found in a range of organisms, from algae, plants and microorganisms to animals (Raemdonck et al. 2013). They can be linear, branched or cyclic and differ in their charges and molecular weights. Advantages of these materials include their non-toxicity, ready biodegradability, eco-friendly aspects and low cost. Usage of non-toxic biodegradable polymers, such as chitin, chitosan, amylose, polyhydroxybutyrate, dextran, agarose, alginate, carrageenan, cellulose and carboxymethyl cellulose,are increasing in agricultural applications (Voinova and Kalacheva 2009) (Fig. 3). These materials have network-like structures which provide controlled release effect and are helpful as compost in fields after their degradation (Kumar et al. 2014). Earlier, fabrication of controlled release materials was exclusively reported with synthetic polymers. Cross linking between polymeric chains forms network-like structures by different interactions like covalent, hydrophobic, hydrogen and van der Waal forces (Dhotel et al. 2013). The main polymers employed in smart delivery of bioactive compounds in agriculture are described below.

Chitosan. Chitosan is a non-toxic, biodegradable, linear copolymer of 2-acetamido-2-deoxy-β-D-glucopyranose and 2-amino-2-deoxy-β-D-glucopyranose. It is produced from chitin by deacetylation using alkaline hydrolysis or enzymatic treatment (Choudhary et al. 2017a). Due to its unique properties, such as biocompatibility, biodegradability, hydrophilicity, safe, and non-toxic nature, chitosan-based nanomaterials are used in several agricultural applications, including antimicrobial activity (Qi et al. 2004, Saharan et al. 2013, 2015), plant growth-promoting activity (Saharan et al. 2015, 2016, Choudhary et al. 2017b) and nano-fertilizers (Corradini et al. 2010, Abdel-Aziz et al. 2016).

Starch. Starch is one of the main reserve food materials in cereal grain, legume and vegetables. It is composed of amylose (linear) and amylopectin (branched) linked by

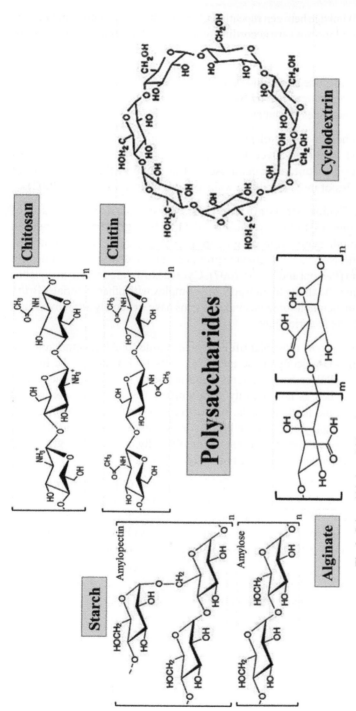

Fig. 3. Selected biodegradable polymers commonly used for smart delivery of agrochemicals.

α-(1,4) and α-(1,6) linkage between repeating glucose units. The relative proportions of amylose and amylopectin vary according to source and developmental stages of plants (Laycock and Halley 2014). In the last decade, starch-based nanomaterials have been in consideration in food science and agriculture due to higher availability, biocompatibility and biodegradability. Starch-based nanomaterials have been used in the making of nanocomposites by blending with bioactive compounds such as detergents and insecticides (Fishman et al. 2000).

Chitin. Chitin is homopolymer of D-glucosamine and N-acetyl-D-glucosamine monomers, joined by β-(1,4) glycosidic bonds (Kobyashi 2007). It is the most abundant polymer in nature after cellulose and is found in the exoskeleton of crustaceans, cuticles of insects and cell walls of fungi (Chandrkrachang 2002).

Cyclodextrins. Cyclodextrins are complex polysaccharides composed of monosaccharide unit, α-D-glucopyranose by α-(1,4) linkage. They are produced from starch through cyclization of glucopyranose by cyclodextrin glucosyl transferase enzyme. Normally this enzyme produces α-cyclodextrin, β-cyclodextrin and γ-cyclodextrin (Pacioni and Veglia 2007). Cyclodextrins are soluble in aqueous suspension and have the ability to transport and complex with other materials (Britto et al. 2004). Due to their non-toxicity, cyclodextrins have extremely attractive smart delivery applications in agriculture.

Cellulose. Cellulose is the most plentiful polysaccharide found in nature and is a linear polymer of β-D-glucopyranose units linked by β-(1,4) glycosidic linkage (Kobyashi 2007). Cellulose-based nanomaterials are being widely used due to their unique properties, such as non-toxicity, biocompatibility, biodegradability and low cost. Cellulose and its derivatives, such as carboxymethyl cellulose (CMC) and hydroxypropyl methylcellulose (HPMC), are widely used as smart delivery systems for bioactive compounds. These are ecofriendly and easily degraded by microorganisms present in the air, soil and water (Akar et al. 2012).

Alginate. Alginate is a linear polysaccharide comprised of β-D-manuronic acid and α-L-guluronic acid joined by 1–4 bonds. It is obtained from brown algae (Lertsutthiwong et al. 2008). Alginate and its derivatives-based formulation have been prepared for a series of applications, including encapsulation of agrochemicals (herbicide, insecticide and bactericide) for controlled and slow release (Singh et al. 2013).

2.3 Chitosan Nanomaterial-Based Delivery System

Natural products are an unbeatable substitute for any synthetic agrochemicals in agriculture to abate the negative impact on the environment. Chitosan, a prominent biopolymer, has sought major attraction for use as a delivery agent due to the combination of its useful structural and biological properties. Chitosan-based nanomaterials have excellent physico-chemical properties due to their smaller size, high surface to volume ratio and surface charge as compared to chitosan

(Saharan et al. 2015). Chitosan-based nanomaterials have been reported innumerous agricultural applications, including seedling growth and development (Saharan et al. 2016), nutrient use efficiency (Corradini et al. 2010, Abdel-Aziz et al. 2016), plant growth (Saharan et al. 2015, Choudhary et al. 2017b), antimicrobial activity (Saharan et al. 2013, 2015), disease control (Saharan et al. 2015, Choudhary et al. 2017b) and elicitors of plant defense mechanisms (Chandra et al. 2015, Sathiyabama and Manikandan 2016, Choudhary et al. 2017b). Synthesis of chitosan-based nanomaterialsis achieved through ionic gelation method, which is based on the interactions between positively charged amino groups of chitosan and negatively charged cross-linking agent, the tri-poly phosphate (TPP) (Fig. 4). Amino group of chitosan nanomaterials also forms complex with other charged polymers, such as poly acrylic acid, carboxymethyl cellulose, xanthan, carrageenan, alginate and pectin (Kashyap et al. 2015). Further, these nanomaterials have also been reported as excellent metal chelators for Cu and Zn (Choudhary et al. 2017a).

A comprehensive range of molecular weights and degrees of acetylation renders chitosan considerably reasonable for the development of nano formulations. In recent years, efforts have been made to develop chitosan nanomaterials as valuable carriers for smart delivery through controlled release of encapsulated agrochemicals, such as macronutrients, micronutrients, pesticide and herbicide, in order to increase the efficiency of large-scale agriculture (Kashyap et al. 2015) (Table 1). The main advantage of encapsulation in chitosan matrix is that it protects the active ingredients from the surrounding environment. Controlled release of encapsulated materials is accomplished due to slow release effect of chitosan nanomaterials and bonding of active ingredients with chitosan. Controlled release system offers advantages, such as use of active ingredients in smaller quantities and reduced loss during leaching, volatilization and run-offs (Campos et al. 2008). The encapsulation of toxic materials, such as pesticide in chitosan polymer, makes it less toxic and more availability to target plants for longer duration.

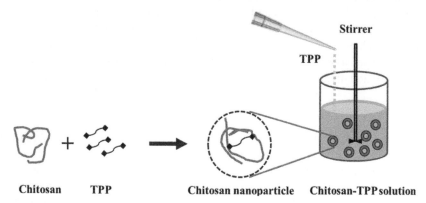

Fig. 4. Preparation of chitosan NPs by ion gelation method.

Table 1. Chitosan nanomaterials for smart delivery of bioactive compounds in agriculture.

Types of Nanoparticle	Application/Purpose	References
Micronutrient Delivery		
Zn-chitosan nanoparticles	Zinc	Deshpande et al. 2017
Cu-Chitosan nanoparticles	Copper	Saharan et al. 2015, 2016, Choudhary et al. 2017b
Fertilizer Delivery		
Chitosan nanoparticles	NPK-fertilizer	Corradini et al. 2010, Abdel Aziz et al. 2016
Pesticide/Biopesticide Delivery		
Chitosan nanoparticles	Etofenprox	Hwang et al. 2011
Chitosan nanoparticles	Acetamiprid	Kumar et al. 2015
Angico Gum/Chitosan Nanoparticles	Lippiasidoides (an essential oil)	Paula et al. 2010
Chitosan nanoparticles	Oregano (an essential oil)	Hosseini et al. 2013
Chitosan/kaolin nanoparticle	Absorption of *Myrothrecium verrucaria* enzyme: A microbial product	Ghormade et al. 2011
Herbicide Delivery		
Chitosan nanoparticles	Paraquat	Grillo et al. 2014, 2015
Chitosan nanoparticles	Imazapic and Imazapyr	Maruyama et al. 2016
Alginate/chitosan nanoparticles	Paraquat	Silva Mdos et al. 2011
Chitosan nanoparticles	Dichlorprop	Wen et al. 2011
Genetic Material Delivery		
Chitosan nanoparticle	Double stranded RNA	Zhang et al. 2010

Chitosan-based nanomaterials have been used for delivery of active ingredients, like insecticides etofenprox (Hwang et al. 2011) and acetamiprid (Kumar et al. 2015), and herbicides dichlorprop (Wen et al. 2011), paraquat (Grillo et al. 2014), imazapic and imazapyr (Maruyama et al. 2016). NPK-fertilizer (Corradini et al. 2010, Abdel Aziz et al. 2016) and micronutrients, such as copper (Saharan et al. 2013, 2015, 2016, Choudhary et al. 2017a,b) and zinc (Choudhary et al. 2017a), have been achieved in chitosan nanoparticles. Furthermore, chitosan nanoparticles have been used as carrier materials for delivery of double stranded RNA (Zhang et al. 2010) and oregano (an essential oil) in plants (Hosseini et al. 2013).

2.3.1 Delivery of Fertilizer

In the post green revolution era, urea consumption increased by 29% which led to an 80% increase in atmospheric N_2O, causing global warming. Currently, urea, diammonium phosphate (DAP) and single superphosphate (SSP) are used as a reservoir of N, P and K in the soil. According to an estimate, an average of 65% of these applied fertilizers is vanished into the environment, causing fiscal loss to the nation (Trankel 1997, Ombodi et al. 2000).

A diverse range of slow and controlled release fertilizers have been developed using synthetic or biopolymers to lessen the potential vulnerability to leaching and denitrification of many nitrogenous fertilizers. The modern approach involves the application of nano coating to urea or other chemical fertilizers. Rate of dissolution of fertilizer is reduced by stability of nano-coating, which permits slow and sustained release of coated fertilizer that is more effectively absorbed by plant roots (Duhan et al. 2017). Chitosan-based encapsulated nano-fertilizer is important in agriculture since it can easily be applied to the leaf surface and enters the stomata, enhancing nutrient uptake and improving plant growth and crop yield. Development of effective tools for controlled and slow release system for delivery of fertilizers at optimum rate to plants has the potential to provide significant cost to the farmers as much of the fertilizer applied to crops is lost due to volatilization, leaching, run-off and other phenomena. Nanocomposites of chitosan with polymethacrylic acid (PMAA) furnished satisfactory results for controlled and slow release of NPK fertilizer (Corradini et al. 2010). In wheat crop, foliar application of chitosan nanoparticles loaded with NPK fertilizer has been reported to enhance various parameters, such as harvest index, crop index and grain yield (Abdel-Aziz et al. 2016). Biodegradable chitosan nanomaterials are favourable for use in controlled release of NPK fertilizer due to their bio-absorbable nature. Their application has been proved safe for germination of cereals. Commercial products of nano-fertilizer having chitosan include "Master Nano Chitosan Organic Fertilizer", containing water soluble liquid chitosan, organic acid and salicylic acids, phenolic compounds manufactured by company Pannaraj Intertrade, Thailand. Some other commercial products of nanofertilizer include Nano-Gro™, Nano Green, Nano-Ag Answer®, Biozar Nano-Fertilizer, Nano Max NPK Fertilizer and TAG NANO (Prasad et al. 2017).

2.3.2 Delivery of Micronutrient

It is well known that micronutrients like manganese, copper, boron, iron, molybdenum and zinc are crucial for the growth and development of plants. Continuous and excessive usage of these micronutrients enhances their concentration in the biosphere and could exceed the specified limit of a particular element in the environment (Deluisa et al. 1996). However, by adopting new farming practices, the concentration of micronutrients in soil can be considerably decreased (Alloway 2009). With the advancement of nano-based delivery systems, the concept of controlled and slow release of micronutrients for stable and long-term effect has emerged. These micronutrients have been targeted to encapsulate into chitosan-based nanomaterials in order to enhance their availability and efficacy in the plants. Copper, iron, silver, zinc and manganese-loaded chitosan NPs have been reported to exhibit antibacterial activity against numerous microorganisms (Du et al. 2004). Cu encapsulated in chitosan nanoparticles showed growth promoting effects in tomato (Saharan et al. 2015). In another study, Cu-chitosan nanoparticles significantly enhanced percent germination, shoot and root length, root number, seedling length, fresh and dry weight and seed vigor index with increased activities of α-amylase and protease enzymes in maize germinating seeds (Saharan et al. 2016). In a recent study, Cu-chitosan nanoparticles were found to enhance antioxidant and defense-related enzymes activities, plant growth, disease control and grain yield (Choudhary et al. 2017b). Deshpande et al. (2017) prepared Zn-complexed chitosan nanoparticles and evaluated for micronutrient nano-carrier. These nanoparticles were found to be suitable as a foliar fertilizer for agronomic bio-fortification for higher grain zinc content in wheat crop. Therefore, chitosan nanocomposite of micronutrient could be an excellent source for improved foliar uptake, including slow release of micronutrient, which positively affects soil health and plant growth.

2.3.3 Delivery of Pesticides/Bio-Pesticides

After World War II, agriculture dependency on synthetic pesticides became widespread. Controlling pest population with minimal impact on environment is the major challenging issue in present agriculture. Mode of application of pesticide greatly influences its efficacy and possible effect on the environment (Ihsan et al. 2007). At present, pesticide application is imparted by knapsacks sprayer or ultra-light volume sprayers that deliver large (9–66 μm) and smaller droplets (3–28 μm) associated with splash loss and spray drift, respectively. Both of these demerits can be effectively overcome by using nanoparticle encapsulated pesticides (Hoffmann et al. 2007).

A noteworthy example of chitosan nanomaterial for delivery of pesticide includes nanochitosan encapsulated beauvericin, prepared by ionic gelation method, which improved pesticidal activity against groundnut defoliator *Spodopteralitura* (Bharani et al. 2014). Several other pesticides, such as etofenprox (Hwang et al. 2011) and acetamiprid (Kumar et al. 2015) have been used along with chitosan nanoparticles for delivery in plants. Chitosan has also been used as a carrier for slow release of active principles, namely rotenone and carvacrol. It is claimed that chitosan-based nanoparticles could be a new area of research for the generation of chitosan-based bionanopesticide.

2.3.4 Delivery of Herbicide

Weeds are the major issues regarding loss to the principal food production, to the extent of 10–15%. This problem leads to an alarming increase in the use of herbicide to manage losses, ultimately resulting in harm to the environment. Bioavailability, chemical stability, solubility, photodegradation and soil sorption and adverse effect on water quality and environment are the major problems associated with herbicides that are currently in use (Kashyap et al. 2015). This can be accomplished successfully via the controlled release phenomenon.

Delivery of paraquat through alginate/chitosan nanoparticles has successfully been done (Silva et al. 2011). In another study, chitosan and tripolyphosphate nanoparticles were prepared as a carrier for paraquat herbicide (Grillo et al. 2014).

Furthermore, herbicides, such as dichlorprop (Wen et al. 2011), paraquat (Grillo et al. 2014), imazapic and imazapyr (Maruyama et al. 2016), were successfully delivered using chitosan nanoparticles in many crop plants. Bio-nanocomposites of chitosan have been used as an adsorbent for herbicides clopyralid and dichlorprop for their controlled release. Chitosan gel beads (with acetic or propionic anhydride) containing atrazine and urea as active ingredients have also been prepared (Teixeira et al. 1990). Limited literature is available on the application of chitosan nanoparticles in herbicide delivery, prompting a demand for research in this area.

2.3.5 Delivery of Genetic Material

In agriculture, delivery of gene is crucial for crop improvement. Traditional methods of gene delivery significantly deviate the plant cell wall. Moreover, these methods have lower (0.01–20%) efficiency of transformation and are not applicable to all type of plants. Therefore, there is a need to develop methods which can significantly overcome these demerits. Nanotechnology provides noteworthy advantages over conventional methods of gene transfer. This approach is applicable to monocot as well as dicot plants and transgene silencing can be overcome by regulating DNA copies via nanoparticles. The significance of using nanochitosan in gene transfer strategy is its protonating ability in acidic solution, allowing it to form complex with DNA, in which positively charged amino groups of chitosan interact electrostatically with negatively charged phosphate group of DNA (Duceppe et al. 2010). Additionally, these complexes are more stable as they protect DNA from nuclease degradation. DNA delivery by chitosan nanoparticle is highly dependent on its molecular weight, degree of deacetylation and pH of the medium, as it affects its binding with DNA and release.

3. Mechanism of Smart Delivery System

In agriculture, smart and targeted delivery of bioactive compounds is needed to minimize the loss caused due to over application and higher concentration and to prevent the harmful side effects of agrochemicals. The encapsulation of active ingredients can be achieved using nanoparticles, nanoemulsions, nanocapsules and micelles. These systems will prevent degradation and allow for the controlled and site-specific release of agrochemicals. Smart delivery of bioactive ingredient is achieved by controlled and slow release properties of the polymeric chitosan nanomaterials, interaction of the ingredients to the nanomaterials and the target environmental conditions (Fig. 5). The release of bioactive ingredients from encapsulated chitosan nanoparticles depends on size, surface charge, density, surface morphology, rate of cross-linking and physico-chemical properties of the ingredient. Further, the release of bioactive ingredients is also affected by pH, solvent property and presence of enzymes under *in vitro* conditions (Kashyap et al. 2015). Generally, release of bioactive ingredients from chitosan-based nanomaterials occurs due to diffusion or degradation release mechanism. The probable mechanisms include

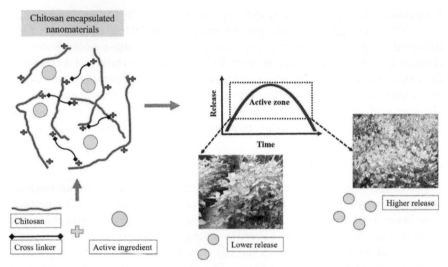

Fig. 5. Representation of a smart delivery system of bioactive compounds based on chitosan nanomaterials in agriculture.

swelling of chitosan matrix by penetration of water, conversion of glassy polymer into swollen matrix and a slow and steady release/diffusion of compounds from the swollen matrix.

Conclusion and Future Prospects

Currently, global agriculture is facing many challenges caused by climate change, uncontrolled use of agrochemicals and nutrient deficiency. The application of such agrochemicals enhances crop production and productivity but their large-scale use is not suitable for the long term. Biopolymer-based nano-delivery systems are an emerging solution for ensuring food safety and sustainability of agricultural production system. Chitosan-based nanomaterials show slow and controlled release of encapsulated materials which can release their active ingredients to the responding environment. Moreover, these nanomaterials can be used for efficient and effective smart delivery of biologically active substances, such as micronutrients, fertilizers, pesticides and genetic material, to the plants. Furthermore, chitosan-based nanomaterials have been successfully used as fertilizer carriers in various crops to enhance nutrient use efficiency and reduce environmental pollutions. The efficacy of these nanomaterials is better than current agrochemicals and synthetic fertilizers. In terms of low cost, low concentration and eco-safety, chitosan nanomaterials may become more common products in agriculture in the near future.

References

Abdel-Aziz, H.M.M., M.N.A. Hasaneen and A.M. Omer. 2016. Nano chitosan-NPK fertilizer enhances the growth and productivity of wheat plants grown in sandy soil. Span. J. Agric. Res. 14(1): e0902.

Abo-Elyousr, K.A.M., M.A.A. Seleim, K.M.H. Abd-El-Moneem and F.A. Saead. 2014. Integrated effect of *Glomus mosseae* and selected plant oils on the control of bacterial wilt disease of tomato. Crop Prot. 66: 67–71.

Akar, E., A. Altınışık and Y. Seki. 2012. Preparation of pH- and ionic-strength responsive biodegradable fumaric acid crosslinked carboxymethyl cellulose. Carbohydr. Polym. 90: 1634–1641.

Alloway, B.J. 2009. Soil factors associated with zinc deficiency in crops and humans. Environ. Geochem. Health 31(5): 537–548.

Anton, N. and T.F. Vandamme. 2011. Nano-emulsions and micro-emulsions: Clarifications of the critical differences. Pharm. Res. 28: 978–985.

Bharani, R.S.A., R.S. Karthick and S.S.S. Namasivayam. 2014. Biocompatible chitosan nanoparticles incorporated pesticidal protein beauvericin (csnp-bv) preparation for the improved pesticidal activity against major groundnut defoliator *Spodopteralitura* (Fab.) (Lepidoptera; Noctuidae). Int. Chem. Tech. Res. 6(12): 5007–5012.

Britto, M.A.F.O., C.S.J. Nascimento and H.F. dos Santos. 2004. Structural analysis of cyclodextrins: A comparative study of classical and quantum mechanical methods. Quim. Nova. 27: 882–888.

Campos, E.V.R., J.L. Oliveira, L.F. Fraceto and B. Singh. 2015. Polysaccharides as safer release systems for agrochemicals. Agron. Sustain. Dev. 35: 47–66.

Carvalho, F.A. 2006. Agriculture, pesticides, food security and food safety. Environ. Sci. Policy 9: 685–692.

Chandra, S., N. Chakarborty, A. Dasgupta, J. Sarkar, K. Panda and K. Acharya. 2015. Chitosan nanoparticle: A positive modulator of innate immune responses in plants. Sci. Rep. 5: 1–13.

Chandrkrachang, S. 2002. The applications of chitin in agriculture in Thailand. Adv. Chitin Sci. 5: 458–462.

Choudhary, R.C., R.V. Kumaraswamy, S. Kumari, S.S. Sharma, A. Pal, R. Raliya, P. Biswas and V. Saharan. 2017a. Cu-chitosan nanoparticle boost defense responses and plant growth in maize (*Zea mays* L.) Sci. Rep. 7: 9754.

Choudhary, R.C., R.V. Kumaraswamy, S. Kumari, S.S. Sharma, A. Pal, R. Raliya, P. Biswas and V. Saharan. 2017b. Synthesis, characterization, and application of chitosan nanomaterials loaded with zinc and copper for plant growth and protection. pp. 227–247. *In*: Prasad, R. et al. (eds.). Nanotechnology. Springer Nature.

Corradini, E., M.R. de Moura and L.H.C. Mattoso. 2010. A preliminary study of the incorporation of NPK fertilizer into chitosan nanoparticles. Express Polym. Lett. 4(8): 509–515.

De, A., R. Bose, A. Kumar and S. Mozumdar. 2014. Targeted delivery of pesticides using biodegradable polymeric nanoparticles, Springer Briefs in Molecular Science. doi:10.1007/978-81-322-1689-6_2.

Deluisa, A., P. Giandon, M. Aichner, P. Bortolami, L. Bruna, A. Lupetti, F. Nardelli and G. Stringari. 1996. Copper pollution in Italian vineyard soils. Commun. Soil Sci. Plant Anal. 27: 1537.

Deshpande, P., A. Dapkekar, M.D. Oak, K.M. Paknikar and J.M. Rajwade. 2017. Zinc complexed chitosan/TPP nanoparticles: A promising micronutrient nanocarrier suited for foliar application. Carbohydr. Polym. 165: 394–401.

Dhotel, A., Z. Chen, L. Delbreilh, B. Youssef, J.M. Saiter and L. Tan. 2013. Molecular motions in functional self-assembled nanostructures. Int. J. Mol. Sci. 14(2): 2303–2333.

Duceppe, N. and M. Tabrizian. 2010. Advances in using chitosan-based nanoparticles for *in vitro* and *in vivo* drug and gene delivery. Expert Opin. Drug Deliv. 7: 1191–1207.

Duhan, J.S., R. Kumar, N. Kumar, P. Kaur, K. Nehra and S. Duhan. 2017. Nanotechnology: The new perspective in precision agriculture. Biotechnol. Rep. 15: 11–23.

Fishman, M.L., D.R. Coffin, R.P. Konstance and C.I. Onwulata. 2000. Extrusion of pectin/starch blends plasticized with glycerol. Carbohydr. Polym. 41: 317–325.

Ghormade, V., M.V. Deshpande and K.M. Paknikar. 2011. Perspectives for nano-biotechnology enabled protection and nutrition of plants. Biotechnol. Adv. 29: 792–803.

Grillo, R., Z. Clemente, J.L. Oliveira, E.V.R. Campos, V.C. Chalupe, C.M. Jonsson, R. Lima, G. Sanches, C.S. Nishisaka, A.H. Rosa, K. Oehlke, R. Greiner and L.F. Fraceto. 2015. Chitosan nanoparticles loaded the herbicide paraquat: The influence of the aquatic humic substances on the colloidal stability and toxicity. J. Hazard Mater. 286: 562–572.

Grillo, R., A.E. Pereira, C.S. Nishisaka, R. de Lima, K. Oehlke, R. Greiner and L.F. Fraceto. 2014. Chitosan/tripolyphosphate nanoparticles loaded with paraquat herbicide: An environmentally safer alternative for weed control. J. Hazard Mater. 27: 163–171.

Gupta, A., H.B. Eral, T.A. Hatton and P.S. Doyle. 2016. Nanoemulsions: Formation, properties and applications. Soft Matter 12: 2826–2841.

Hoffmann, W.C., T.W. Walker, V.I. Smith, D.E. Martin and B.K. Fritz. 2007. Droplet-size characterization of handheld atomization equipment typically used in vector control. J. Am. Mosq. Control Assoc. 23: 315–320.

Hosseini, S.F., M. Zandi, M. Rezaei and F. Farahmandghavi. 2013. Two-step method for encapsulation of oregano essential oil in chitosan nanoparticles: Preparation, characterization and *in vitro* release study. Carbohydr. Polym. 95(1): 50–6.

Hwang, I.C., T.H. Kim, S.H. Bang, K.S. Kim, H.R. Kwon, M.J. Seo, Y.N. Youn, H.J. Park, C. Yasunaga-Aoki and Y.M. Yu. 2011. Insecticidal effect of controlled release formulations of etofenprox based on nano-bio technique. J. Fac. Agr. Kyushu U. 56: 33–40.

Ihsan, M., A. Mahmood, M.A. Mian and N.M. Cheema. 2007. Effect of different methods of fertilizer application to wheat after germination under rainfed conditions. J. Agric. Res. 45: 277–281.

Kashyap, P.L., X. Xiang and P. Heiden. 2015. Chitosan nanoparticle-based delivery systems for sustainable agriculture. Int. J. Biol. Macromol. 77: 36–51.

Kobyashi, B.S. 2007. New developments of polysaccharide synthesis via enzymatic polymerization. Proc. Jpn. Acad. Ser. B 83: 215–246.

Kohler, H.R. and R. Triebskorn. 2013. Wildlife ecotoxicology of pesticides: Can we track effects to the population level and beyond? Sci. 341: 759–65.

Kumar, S., N. Chauhan, M. Gopal, R. Kumar and N. Dilbaghi. 2015. Development and evaluation of alginate-chitosan nanocapsules for controlled release of acetamiprid. Int. J. Biol. Macromol. 81: 631–637.

Kumaraswamy, R.V., S. Kumari, R.C. Choudhary, A. Pal, R. Raliya, P. Biswas and V. Saharan. 2018. Engineered chitosan based nanomaterials: Bioactivities,mechanisms and perspectives in plant protection and growth. Int. J. Biol. Macromol. 113: 494–506.

Kumari, S., R.V. Kumaraswamy, R.C. Choudhary, S.S. Sharma, A. Pal, R. Raliya, P. Biswas and V. Saharan. 2018. Thymol nanoemulsion exhibits potential antibacterial activity against bacterial pustule disease and growth promotory effect on soybean. Sci. Rep. 8: 6650.

Lao, S.B., Z.X. Zhang, H.H. Xu and G.B. Jiang. 2010. Novel amphiphilic chitosan derivatives: Synthesis, characterization and micellar solubilization of rotenone. Carbohydr. Polym. 82: 1136–1142.

Laycock, B.G. and P.J. Halley. 2014. Starch applications: State of market and new trends. pp. 381–419. *In*: Avérous, P.J.H. (ed.). Starch Polymer. Elsevier, Amsterdam.

Lertsutthiwong, P., K. Noomun, N. Jongaroonngamsang, P. Rojsitthisak and U. Nimmannit. 2008. Preparation of alginate nanocapsules containing turmeric oil. Carbohydr. Polym. 74: 209–214.

Maruyama, C.R., M. Guilger, M. Pascoli, N. Bileshy-Jose, P.C. Abhilash, L.F. Fraceto and R. de Lima. 2016. Nanoparticles based on chitosan as carriers for the combined herbicides imazapic and imazapyr. Sci. Rep. 6: 23854.

Mason, T.G., J.N. Wilking, K. Meleson, C.B. Chang and S.M. Graves. 2006. Nanoemulsions: Formation, structure and physical properties. J. Phys. Condens. Matter 18(41): 635–666.

Nair, R., S.H. Varghese, B.G. Nair, T. Maekawa, Y. Yoshida and D.S. Kumar. 2010. Nanoparticulate material delivery to plants. Plant Sci. 179: 154–163.

Ombodi, A. and M. Saigusa. 2000. Broadcast application versus band application of polyolefin coated fertilizer on green peppers grown on andisol. J. Plant Nutr. 23: 1485–1493.

Pacioni, N.L. and A.V. Veglia. 2007. Determination of poorly fluorescent carbamate pesticides in water, bendiocarb and promecarb, using cyclodextrin nanocavities and related media. Anal. Chim. Acta. 583: 63–71.

Parisi, C., M. Vigani and E.R. Cerezo. 2015. Agricultural nanotechnologies: What are the current possibilities? Nano Today 10: 124–127.

Paula, H.C.B., F.M. Sombra, F.O.M.S. Abreu and R.C.M. de Paula. 2010. Lippiasidoides essential oil encapsulation by angico gum/chitosan nanoparticles. J. Braz. Chem. Soc. 21: 2359–2366.

Prasad, R., A. Bhattacharyya and Q.D. Naguyen. 2017. Nanotechnology in sustainable agriculture: Recent developments, challenges, and perspectives. Front. Microbiol. 8: 1014.

Qi, L., Z. Xu, X. Jiang, C. Hu and X. Zou. 2004. Preparation and antibacterial activity of chitosan nanoparticles. Carbohydr. Res. 339: 2693–2700.

Raemdonck, K., T.F. Martens, K. Braeckmans, J. Demeester and S.C. De Smedt. 2013. Polysaccharide based nucleic acid nanoformulations. Adv. Drug Deliv. Rev. 65: 1123–1147.

Rai, V., S. Acharya and N. Dey. 2012. Implications of nanobiosensors in agriculture. J. Biomater. Nanobiotchnol. 3: 315–324.

Saharan, V., R.V. Kumaraswamy, R.C. Choudhary, S. Kumari, A. Pal, R. Raliya and P. Biswas. 2016. Cu-chitosan nanoparticle mediated sustainable approach to enhance seedling growth in maize by mobilizing reserved food. J. Agric. Food Chem. 64(31): 6148–6155.

Saharan, V., A. Mehrotra, R. Khatik, P. Rawal, S.S. Sharma and A. Pal. 2013. Synthesis of chitosan based nanoparticles and their *in vitro* evaluation against phytopathogenic fungi. Int. J. Biol. Macromol. 62: 677–683.

Saharan, V., G. Sharma, M. Yadav, M.K. Choudhary, S.S. Sharma, A. Pal, R. Raliya and P. Biswas. 2015. Synthesis and *in vitro* antifungal efficacy of Cu-chitosan nanoparticles against pathogenic fungi of tomato. Int. J. Biol. Macromol. 75: 346–353.

Sathiyabama, M. and A. Manikandan. 2016. Chitosan nanoparticle induced defense responses in finger millet plants against blast disease caused by *Pyriculariagrisea* (Cke.) Sacc. Carbohydr. Polym. 154: 241–246.

Silva Mdos, S., D.S. Cocenza, R. Grillo, N.F. de Melo, P.S. Tonello, L.C. de Oliveira, D.L. Cassimiro, A.H. Rosa and L.F. Fraceto. 2011. Paraquat-loaded alginate/chitosan nanoparticles: Preparation, characterization and soil sorption studies. J. Hazard. Mater. 190: 366–374.

Singh, B., D.K. Sharma and A. Dhiman. 2013. Environment friendly agar and alginate-based thiram delivery system. Toxicol. Environ. Chem. 95: 567–578.

Singh, T., S. Shukla, P. Kumar, V. Wahla, V.K. Bajpai and I.A. Rather. 2017. Application of nanotechnology in food science: Perception and overview. Front. Microbiol. 8: 1501.

Tahat, M.M., S. Kamaruzaman and O. Radziah. 2011. Bio-compartmental *in vitro* system for *Glomus mosseae* and *Ralstoniasolanacraum* interaction. Inter. J. Bot. 7: 295–299.

Tarafdar, J.C., S. Sharma and R. Raliya. 2013. Nanotechnology: Interdisciplinary science of applications. Afr. J. Biotech. 12(3): 219–226.

Teixeira, M.A., W.J. Paterson, E.J. Dunn, Q. Li, B.K. Hunter and M.F.A. Goosen. 1990. Assessment of chitosan gels for the controlled release of agrochemicals. Ind. Eng. Chem. Res. 29: 1205–1209.

Thornton, P.K. 2010. Livestock production: Recent trends, future prospects. Phil. Trans. R. Soc. B 365(1554): 2853–2867.

Tiyaboonchai, W. 2003. Chitosan nanoparticles: A promising system for drug delivery. Naresuan Univ. J. 11: 51–66.

Trenkel, M.E. 1997. Controlled-release and stabilized fertilizers in agriculture. Int. Fertilizer Industry Association Paris, pp. 234–318.

Tsuji, K. 2001. Microencapsulation of pesticides and their improved handling safety. J. Microencapsul. 18: 137–47.

Voinova, O.N., G.S. Kalacheva, I.D. Grodnitskaya and T.G. Volova. 2009. Microbial polymers as a degradable carrier for pesticide delivery. Appl. Biochem. Microbiol. 45: 384–388.

Wang, X., G. Huang and J. Liu. 2014. Projected increases in near-surface air temperature over Ontario, Canada: A regional climate modeling approach. Clim. Dyn. doi:10.1007/s00382-014-2387-y.

Wen, Y., H. Chen, Y. Yuan, D. Xu and X. Kang. 2011. Enantioselective ecotoxicity of the herbicide dichlorprop and complexes formed with chitosan in two fresh water green algae. J. Environ. Monitor JEM 13: 879–885.

Zhang, X., J. Zhang and K.Y. Zhu. 2010. Chitosan/double-stranded RNA nanoparticle-mediated RNA interference to silence chitin synthase genes through larval feeding in the African malaria mosquito (*Anopheles gambiae*). Insect Mol. Biol. 19: 683–693.

CHAPTER 8

Interaction of Nanoparticles with Crop Plants

Their Implication in Nutritional Imbalance and Environmental Toxicity

Samarendra Barik,[1,#] *Saheli Pradhan*[1,#,*] and
Arunava Goswami[1]

1. Introduction

Nutrients, called essential elements, are required by all plants in order to complete their life cycle, since each essential element has a direct effect on plant growth and reproduction. In the excess of any of the elements, or its severe deficiency, the plant will die before it completes its life cycle. This requirement acknowledges the fact that the elements have a function in metabolism that, with their short supply, may develop abnormal symptoms of deficiency. As a consequence of disrupted metabolism, plants may sometimes able to complete their life cycles but often with restricted growth and abnormal appearance. Hence, these elements are unique in plant metabolism and no other substrate can totally substitute their functionality.

Besides carbon (C), hydrogen (H) and oxygen (O), which form the basic skeleton of a plant, other elements like nitrogen (N), potassium (K), phosphorus (P), calcium (Ca), magnesium (Mg), sulfur (S), iron (Fe), manganese (Mn), copper (Cu), boron (B), zinc (Zn), molybdenum (Mo), chlorine (Cl) and nickel (Ni). Sodium (Na), silicon (Si) and cobalt (Co) are all required for a plant's growth and development, though many of them are species specific (Taiz and Zeiger 1998). An inadequate availability of such essential elements manifests itself as characteristic deficiency symptoms in the plant. Symptoms may appear in various forms: Yellowing of the

[1] Biological Sciences Division, Indian Statistical Institute, 203 B.T. Road, Kolkata 700108, India.
[#] Contributed equally
[*] Corresponding author: saheli.pradhan@gmail.com

green tissue, called chlorosis, which may be uniform or interveinal in appearance. Necrosis is the after effect of nutritional dysfunction and may occur at leaf tip or marginal and/or interveinal regions of the leaf. Significant inhibition of growth or stunting in appearance with normal green color or off-green/yellow coloration may appear as a result of nutrient deficiency. Excess accumulation of these elements is also detrimental to a plant and leads to toxicity symptoms, as shown in Fig. 1.

When nutritional homeostasis within the plant system is disrupted it leads to a physiological imbalance within the tissue. Diagnosis of these problems can be complicated as deficiency of several elements may occur simultaneously in different plant tissues. Also, some viral plant diseases show symptoms similar to those caused by nutritional deficiency or toxicity (Barker and Pilbeam 2007). Therefore, detailed studies on the nutritional status of the plants are indispensable, as high agricultural yields depend heavily on fertilization with mineral nutrient supplements. In fact, the yield of most crop plants increases in a direct proportion to the amount of the fertilizer they absorb. Because of the poorly maintained irrigation system and low level of physiologically available nutrients, growth and yield of crop plants have been decreasing alarmingly (Barker and Pilbeam 2007).

New technologies need to be introduced in this regard, offering efficient and cost-effective means to increase crop productivity and diverse value-added products. Agricultural innovation has always involved new science-based products and processes that have contributed reliable methods for increasing agricultural productivity and sustainability. One of the newer technologies that are being implemented for crop improvement is biotechnology to modify the genomic structure of the plants. The other emerging technology is the effective implementation of nanotechnology-based plant nutrient supply. In fact, both biotechnology and nanotechnology are complementary as the newer genetic

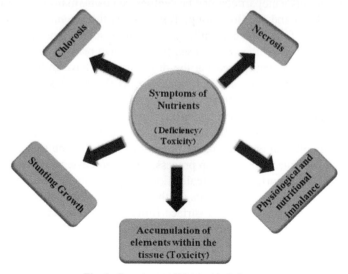

Fig. 1. Symptoms of Nutrient imbalance.

make-up will require more and more efficient nutrient utilization. Preparedness for effective use of nanotechnology is an essential as that for biotechnology.

2. Nanotechnology: A New Era of Science

Nanotechnology is the new era of science which explores the investigation of materials typically in the size range of 10^{-9} meter. Property of these nanomaterials is governed by both classical physics/chemistry and quantum mechanics. This new science was pioneered by American physicist Richard Feynman in his famous lecture "There's Plenty of Room at the Bottom". Since then, nanotechnology has taken a long stride into the modern research field. Presently, overwhelming applications of nanotechnology have expanded to catalysis, optical communications, sensing, optoelectronics and bio-medical sectors (Klaine et al. 2008, Zhao and Castranova 2011). Normally nanomaterials possess small size, high surface area, porosity and high surface area per volume ratio which results in alterable chemical and physical properties compared to its bulk materials. Surface functionality, magnetic properties, aggregation properties and often unique emission properties contribute to their improved catalytic properties (Salata 2004). Depending upon the localization of the electrons, nanomaterials are classified as: (a) metallic/conductor (have delocalized electrons) or (b) insulator (have localized electrons). In terms of quantum mechanics, often there is a restriction of electronic movement within the system. Considering such restriction of electronic movement, nanomaterials are classified as: (a) zero-dimensional (quantum dots), (b) 1-dimensional (nanowires), (c) 2-dimensional (nanotubes) and (d) 3-dimensional (clusters). Band-gap (a difference between the valence band and conductance band) plays the pivotal role of such electronic movement.

Nanomaterials/nanoparticles can be synthesized using mainly two processes, viz. (a) Top-down and (b) Bottom-up. Top-down is the mechanical process for synthesizing NPs. Grinding, ball-milling, very strong sonications are some of the common methods to synthesize NPs via the top-down process. It is noteworthy to mention that NPs synthesized via top-down process have in-homogeneity in sizes and shapes. On the other hand, bottom-up is the chemical process for the synthesis of NPs, where one can start with a certain precursor and using chemical route can end up producing NPs. Flame pyrolysis, high-temperature evaporation, plasma synthesis, microwave radiation, physical and chemical vapor deposition synthesis and colloidal or liquid phase method are some of the reliable methods to synthesize NPs via the bottom-up process. These processes are summarized in Fig. 2. Sometimes in surface functionalization, a capping agent is required in order to produce a stable dispersion of NPs synthesized via the wet chemical route. In a nutshell, bottom-up method and surface functionalization play the pivotal for producing stable NPs and this method is now being used widely for the production of NPs.

The synthesized nanoparticles (NP) have to be surface modified in order to passivate and stabilize, since their nanoscale renders them to be chemically reactive

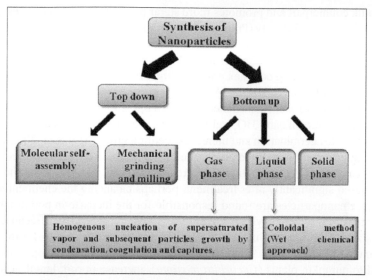

Fig. 2. Synthesis of NPs.

and/or physically aggregative. NPs are also surface functionalized in order to meet the needs of the specific application.

So far, application of NPs in agriculture is at a nascent stage though many studies have been initiated in order to combat nutritional deficiency, toxicity issues and smart gene delivery (Himmelweit 1960, González-Melendi et al. 2008, Pradhan et al. 2011) by the development of new functional materials for enhanced crop productivity, food safety and bio-security (Chinnamuthu and Boopathi 2009, Moraru et al. 2003).

3. Nanoparticles (NPs) on Plant System: NPs on Plant Growth and Physiology

Several works have been performed in order to assess toxicological response to the plants. Some of the NPs are found to be effective in plant system because of their modulation in biochemical and physiological responses. In this context, it is worth mentioning that nano titanium dioxide (TiO_2) is found to enhance the nitrogen metabolism in spinach. As a result, there has been an increase in both fresh and dry weight as an after effect (Yang et al. 2006). Zheng et al. (2005) have also shown that nano TiO_2 encourages the germination as well as growth stage of plants, ensuring the increase in plant dry weight and chlorophyll content. It strengthens the basic question of whether NP can regulate the efficiency of photosynthetic activity as well as photochemical events in plants. Manganese nanoparticles (MnNP) have also been found to enhance the growth and productivity of mung bean plants by modulating the light harvesting process and light-mediated photochemical reactions. After 15 days of treatment from germination, it surpasses the toxic effect of its bulk counterpart and promotes more sugar production, which is an essential criterion of crop productivity (Pradhan et al. 2013); this is because of the fine tuning of MnNP in the process of electron transfer chain (ETC) in the light reaction of photosynthesis. The association of MnNP with light-harvesting complex (LHC) as well as water splitting complex enhances light reaction and also the subsequent dark reaction of photosynthesis. As a consequence, more sugar, which is the ultimate product for plant sustenance and development, is likely being produced. Similarly, a mixture of nano TiO$_2$ and nano silica (SiO$_2$), when treated with soybean seeds, promotes germination and increases seedling growth (Lu et al. 2002). Plant protecting-enzymes of the anti oxidative defense system, like superoxide dismutase (SOD), catalase (CAT) and glutathione reductase (GR) content, have shown to be increased significantly after treatment, perhaps balancing the chemical stress. Magnetite nanoparticles are found responsible for the increase in pod number as well as the dry weight of soybean leaves (Sheykhbaglou et al. 2010). Because of its small size, it facilitates the uptake of iron during photosynthate transfer to leaves of peanut (Liu et al. 2010, Wang et al. 2013). When magnetic nanoparticles are treated to pumpkin plants (*Cucurbita maxima*), increase in root length has been observed (Feizi et al. 2013). On the other hand, Feizi et al. have shown that nano TiO$_2$ at 60 ppm enhances seed germination rate in fennel (*Foeniculum vulgare* Mill) and can be used to improve seed germination in fennel plants (Lin and Xing 2007). In a detailed experiment, they have evaluated the phytotoxicity of five NPs (multi-walled carbon nanotubes (MWCNT), aluminum (Al), alumina (Al$_2$O$_3$), zinc (Zn) and zinc oxide (ZnO) nanoparticles), on seed germination and root growth of radish, rape, rye-grass, lettuce, corn and cucumber (Lee et al. 2010). In that case, only ZnNP and ZnONP are found to be a significant inhibitor of seed germination and root growth. Another metal nanoparticle such as iron nanoparticles (FeNP), when applied on *Brassica juncea*, brought about changes in the level of pigments and other biochemical events. This might be a case for adaptation to environmental stresses within the plant system (Parsons et al. 2010). Surface modified nickel hydroxide NPs, however, did not impart any toxic effect under the experimental condition in seedling stage of mesquite (*Prospis* sp.) (Doshi et al. 2008). Similarly, Aluminum NPs also failed to show toxic symptoms or affect the growth of red kidney bean plants at the dosage of 10–1000 mg/kg with respect to untreated one (Tan and Fugetsu 2007).

Many efforts have been made to implicate baleful action of NPs in plants. Most of these studies have been focused on the stress physiology. Though cytotoxic as well as genotoxic assessments have also been carried out. Tan and Fugetsu (2009) have shown that multi-walled carbon nanotubes (MWCNT), when applied in low doses, decimate cultured rice cell suspension. A probable reason is supposed to be necrosis but, at the higher dose, it causes cell apoptosis. Later on, single walled carbon nanotube (SWCNT) also has been found to have potential cytotoxicity as it results in abundant endonucleolytic cleavage of DNA in *Arabidopsis* cells (Shen et al. 2010). Silver nanoparticles (AgNP) are known to have mitodepressive and mitoclassic effect when treated with meristematic cells of *Allium cepa* root

(Kumari et al. 2009, Babu et al. 2008). There has been a sharp decline in mitotic index in AgNP treated *A. cepa* cells, from 60.3% (untreated one) to 27.62% (at a concentration of 100 mg/L). The main reason behind this downfall is perhaps the inhibition of DNA synthesis at S-phase of the cell cycle. Under similar category, Kumari et al. (2008) have shown that AgNP disturbs the metaphase stage of the cell cycle as chromosomal aberration like stickiness, breakage or gap may be seen during NP treatments. These observations have further been confirmed by Babu et al. through *in vivo* experiments, in which mitotic aberrations like C-metaphase, sticky bridge, laggard chromosome, anaphase-bridge and disturbed anaphase were observed under AgNP treatment (Gardea-Torresdey et al. 2002). Therefore, some NPs can damage the cellular framework not only by inducing oxidative stress but also by influencing genotoxicity and/or cytotoxicity. Nevertheless, many NPs are found to ameliorate plant growth and physiology. Biosafety study of the application of NPs in any biological system is imperative since some of the NPs do affect genomic framework.

4. Nanoparticles (NPs) on Plant System: Biotransformation of NP in Plants

Biotransformation of NP from its ionic state in the plant tissue has not been studied adequately as only a few reports are available. Gardea-Torresday has shown that Au (III) and Ag (I) can reduce the agar growth media accumulated inside the alfalfa (*Medicago sativa*) seedling under experimental conditions (Bali et al. 2010). A similar result has also been observed with Pt (II), where platinum ion accumulates inside the mustard seedling in the form of PtNP. It ensures the fact that NPs can be produced or/and accumulated in the plant in a controlled environment (Harris and Bali 2008, Lopez-Moreno et al. 2010). In the case of ZnO NP, Zn^{2+} ionic state of Zn has been found inside the soybean plant (Gardea-Torresdey et al. 2003). The transformation from ZnO to Zn^{2+} occurs within the root tissue, but in the case of nano CeO_2, there is no transformation inside the root tissue when applied in alfalfa and tomato plants. Surprisingly, NPs are absorbed by the root but not translocated or transformed to ionic state in leaf tissue (Gardea-Torresdey et al. 2003, Birbaum et al. 2010). Biotransformation of NP to ionic state is often found to be plant-specific and depends on the nature of the chemicals applied to the plants. This is a very new field of research but highly significant in terms of understanding the mode of action of NP inside the plant cells and after-effect of their transformation by the cellular metabolism.

5. Nanoparticles (NPs) on Plant System: Uptake, Translocation, and Accumulation of NP Inside the Plants

Plants can absorb and translocate ions through different mechanistic pathways like apoplastic, symplastic or transmembrane pathways or ever through aquaporins. Water-soluble NPs can move through the continuous system of the cell wall, intercellular air spaces or any water-filled extracellular spaces when they travel across the root cortex. They may also cross an entire network of cell cytoplasm through plasmodesmata. Because of water movement in both apoplastic and symplastic pathway, NPs do not necessarily have to cross any semi-permeable membrane; the relevant driving force for mass flow is the gradient in hydrostatic pressure. However, when NPs move across the tonoplast, transportation occurs through the transmembrane pathway. All these processes facilitate the movement of NPs within the cellular network so that plants can readily translocate them from root to leaves in order to perform physiological functions after absorption from the rhizosphere. There has been a lot of effort to understand the dissolution of NP compared to its bulk counterpart (Liu 2009, Birbaum et al. 2010). Bulk counterpart sometimes could undergo solubilization readily, but stable NPs could not be solubilized that easily. The medium components and pH of soil play a pivotal role in dissolution. Therefore, partial dissolution of NPs at acidic pH cannot be ruled out. A negligible quantity of dissolved NPs to yield solubilized minerals can affect these processes as well (Canas et al. 2008). It is obvious that because of the large surface area of roots and their ability to absorb ions at low concentrations from soil solution, plants can easily absorb minerals; but the question arises that, after its absorption, what are their fate and how do they leach out or transform to ionic form so that they can modulate the physiological as well as biochemical pathways of the cellular system. However, the mode of action of NP mostly depends on the size, surface functionality, stability and chemical properties of the particles.

Water-soluble SWCNT can penetrate the intact cell wall through a fluidic phase of endocytosis in *Arabidopsis thaliana*, though water insoluble NP, due to its large size, cannot penetrate the cell wall. In the case of MWCNT, seedlings of tomato can readily absorb by creating new pores in the seed coat, thereby enhancing water uptake and ultimately piercing the epidermal and root hair cell walls and rootcaps of the tissues. Canas et al. 2008 have shown that the uptake kinetics and accumulation of NP are not only size-dependant but also invariably influenced by the surface functionality of the particles. There was no uptake of SWCNT but, after its surface functionalization, nutrient absorption is expedited within 48 h of treatment. Presence of NPs has been observed on the external surface of the primary and secondary roots in the form of nanotube sheets when SWCNT was functionalized with the organic counterpart. Racuciu et al. (2009) have also shown that water-based magnetite NPs coated with perchloric acid decrease nucleic acid content in corn cells, but after surface modification with β-cyclodextrin increase mitotic index (MI) with increasing volume fraction of magnetite NPs (Hyung et al. 2007). NPs have also been conjugated with natural organic matter (NOM) which is a collection of heterogeneous organic substances from decomposed living species. NOM-C_{70} nanocomposite can effectively transmit water and nutrients through conducting tubes to leaves but insignificant uptake has been observed in seed after treatment (Chen et al. 2010). Another important observation has been encountered when NOM was conjugated with both fullerene and fullerols in *Allium cepa*. Uptake of NOM-fullerene nanocomposite is prevented by the plant due to blockage of membrane pores in aleurone layer, however, small sized fullerol conjugated with NOM has eventually been taken up by the plant via apoplastic pathway (Kurepa et al. 2010). Similarly, small sized nano TiO_2 (< 5 nm) complexed with alizarin-S has been taken up by *Arabidopsis thaliana* seedling following tissue and cell-specific distribution. The root of *A. thaliana* secretes mucilage that formed a pectin hydrogel capsule surrounding the root facilitating entry of NPs (Stampoulis et al. 2009).

Hence, surface modification is imperative in NP synthesis as it may either passivate a ver y reactive NP or stabilize a very aggregative NP in the medium where NPs are to be dispersed. It may promote the assembly of NPs or can be under auspices of molecular recognition. Presence of a functional group reduces the free energy of NP surface. For NPs with the high energy of bare surface, the reduction may be vast and restive. Surface functionalization is always welcomed as it lowers adhesion which, in turn, improves uptake level but in a regulated manner. That is the probable reason why Ag concentration in AgNP treated *Cucurbita pepo* plants was found to be almost 4.7 times higher in shoot compared to its bulk counterpart in similar concentration (Stampoulis et al. 2009).

NPs are small in size compared to their bulk materials and possess a very high surface area and often porosity. Typically, surface area per volume ratio is higher in NPs, which results in higher reactivity of NPs compared to their bulk counterpart. The more the surface area and reactivity, the more will be the interaction possibility with the biological network. This could ultimately lead to the unique response within the biological domain. Although the NPs are mostly stable within the biological network, yet a few of them undergo partial dissolution. This phenomenon is indeed the result of a pH deviation which is often observed within a biological network in order to maintain normal biological processes (Franklin et al. 2007). Thus, a pH deviation, especially the acidic pH, could influence the partial dissolution of the NPs, resulting in the formation of corresponding ionic species within the biological regime. According to Franklin et al. (2007) among ZnO NP, bulk ZnO and zinc chloride salt, the toxicity of ZnO NP and bulk ZnO are encountered in *Pseudokirchneriella subcapitata* solely due to dissolved Zn.

Indeed, studies on uptake, translocation and its accumulation of NP in the plant system are crucial in relating interaction of NP with living system, especially from the point of view of their safe implementation in the biological system. Many smart nanomaterials have been designed in order to mitigate the nutritional deficiency, but more work needs to be done to combat the problems of nutritional disorders and declined crop productivity. The efficacy of NP is implicitly dependant on particle size, temperature, chemical reactivity, stability, surface functionalization and plant species (Gao et al. 2006, Linglan et al. 2008, Yang et al. 2007). Therefore, by manipulating these variable factors, smart tools can be synthesized which can negate the basic problems in agriculture, associated with crop productivity and sustainability, through safe ecotoxicological assessment.

Acknowledgements

This work was generously supported by major grants from the Department of Biotechnology, Government of India (DBT) (Grant No: BT/518/NE/TBP/2013), Department of Science and Technology (DST) (Grant No: SR/NM/NS-08/2014), Nanotechnology platform-ICAR (E005) and ISI plan project funds (2014–2017) for providing financial support.

References

Babu, K., M. Deepa, S.G. Shankar and S. Rai. 2008. Effect of nano-silver on cell division and mitotic chromosomes: A prefatory siren. Journal of International Nanotechnology 2: 25–29.

Bali, R., R. Siegele and A.T. Harris. 2010. Biogenic Pt uptake and nanoparticle formation in *Medicago sativa* and *Brassica juncea*. Journal of Nanoparticle Research 12: 3087–3095.

Barker, A.V. and D.J. Pilbeam. 2007. Handbook of Plant Physiology, CRC Press, Taylor and Francis Group.

Bernhardt, E.S., B.P. Colman, M.F. Hochella Jr., B.J. Cardinale and R.M. Nisbet. 2010. An ecological perspective on nanomaterial impacts in the environment. Journal of Environmental Quality 39: 1954–1965.

Birbaum, K., R. Brogiolli, M. Schellenberg, E. Martinoia, W.J. Stark, D. Gunther and L. Limbach. 2010. No evidence for cerium dioxide nanoparticle translocation in maize plants. Environmental Science and Technology 44: 8718–8723.

Canas, J.E., M. Long, S. Nations, R. Vadan, L. Dai, M. Luo, R. Ambikapathi, E.H. Lee and D. Olszyk. 2008. Effects of functionalized and nonfunctionalized single-walled carbon-nanotubes on root elongation of select crop species. Nanomaterials and the Environment 27: 1922–1931.

Chen, R., T.A. Ratnikova, M.B. Stone, S. Lin, M. Lard, G. Huang, J.S. Hudson and P.C. Ke. 2010. Differential uptake of carbon nanoparticles by plant and mammalian cells. Small 6: 612–617.

Chinnamuthu, C.R. and P.M. Boopathi. 2009. Nanotechnology and agroecosystem. Madras Agricultural Journal 96: 17–31.

Doshi, R., W. Braida, C. Christodoulatos, M. Wazne and G. O'Connor. 2008. Nano-aluminum: Transport through sand columns and environmental effects on plants and soil communities. Environmental Research 106: 296–303.

Feizi, H., M. Kamali, L. Jafari and P.R. Moghaddam. 2013. Phytotoxicity and stimulatory impacts of nanosized and bulk titanium dioxide on fennel (Foeniculum vulgare Mill). Chemosphere 91: 506–511.

Franklin, N.M., N.J. Rogers, S.C. Apte, G.E. Batley, G.E. Gadd and P.S. Casey. 2007. Comparative toxicity of nanoparticulate ZnO, bulk ZnO, and ZnCl$_2$ to a freshwater microalga (*Pseudokirchneriella subcapitata*): The importance of particle solubility. Environmental Science and Technology 41: 8484–8490.

Gao, F., F. Hong, C. Liu, L. Zheng, M. Su, X. Wu, F. Yang, C. Wu and P. Yang. 2006. Mechanism of nano-anatase TiO$_2$ on promoting photosynthetic carbon reaction of spinach. Biological Trace Element Research 111: 239–253.

Gardea-Torresdey, J.L., E. Gomez, J. Peralta-Videa, J.G. Parsons, H.E. Troiani and M.J. Yacaman. 2003. Alfalfa sprouts: A natural source for the synthesis of silver nanoparticles. Langmuir 19: 1357–1361.

Gardea-Torresdey, J.L., J.G. Parsons, E. Gomez, J. Peralta-Videa, H.E. Troiani, P. Santiago and M.J. Yacaman. 2002. Formation and growth of Au nanoparticles inside live alfalfa plants. Nano Letters 2: 397–401.

González-Melendi, P., R. Fernández-Pacheco, M.J. Coronado, E. Corredor and P.S. Testillano. 2008. Nanoparticles as smart treatment-delivery systems in plants: Assessment of different techniques of microscopy for their visualization in plant tissues. Annals of Botany 101: 187–195.

Harris, A.T. and R. Bali. 2008. On the formation and extent of uptake of silver nanoparticles by live plants. Journal of Nanoparticle Research 10: 691–695.

Himmelweit, F. 1960. The collected papers of Paul Ehrlich. London: Pergamon Press.

Hyung, H., J.D. Fortner, J.B. Hughes and K.J. Hong. 2007. Natural organic matter stabilizes carbon nanotubes in the aqueous phase. Environmental Science and Technology 41: 179–184.

Klaine, S.J., P.J.J. Alvarez, G.E. Batley, T.F. Fernandes, R.D. Handy, D.Y. Lyon, S. Mahendra, M.J. McLaughlin and J.R. Lead. 2008. Nanomaterials in the environment: Behavior, fate, bioavailability, and effects. Environmental Toxicology and Chemistry 27: 1825–1851.

Kumari, M., A. Mukherjee and N. Chadrasekaran. 2009. Genotoxicity of silver nanoparticle in *Allium cepa*. Science of Total Environment 407: 5243–5246.

Kurepa, J., T. Paunesku, S. Vogt, H. Arora, B.M. Rabatic, J. Lu, M.B. Wanzer, G.E. Woloschak and J.A Smalle. 2010. Uptake and distribution of ultra-small anatase TiO_2 alizarin red S nanoconjugates in Arabidopsis thaliana. Nano Letters 14: 2296–2302.

Lee, C.W., S. Mahendra, K. Zodrow, D. Li, Y.C. Tsai, J. Braam and P.J.J. Alvarez. 2010. Developmental phytotoxicity of metal oxide nanoparticles to *Arabidopsis thaliana*. Environmental Toxicology and Chemistry 29: 669–675.

Lin, D. and B. Xing. 2007. Phytotoxicity of nanoparticles: Inhibition of seed germination and root growth. Environmental Pollution 150: 243–250.

Linglan, M., L. Chao, Q. Chunxiang, Y. Sitao, L. Jie, G. Fengqing and H. Fashui. 2008. Rubisco activase mRNA expression in spinach: Modulation by nanoanatase treatment. Biological Trace Element Research 122: 168–178.

Liu, X.M., F.D. Zhang, S.Q. Zhang, X.S. He, R. Fang, Z. Feng and Y. Wang. 2010. Effects of nano-ferric oxide on the growth and nutrients absorption of peanut. Plant Nutrition and Fertilizer Science 11: 14–18.

Liu, Q., B. Chen, Q. Wang, X. Shi, Z. Xiao, J. Lin and X. Fang. 2009. Carbon nanotubes as molecular transporters for walled plant cells. Nano Letters 9: 1007–1010.

Lopez-Moreno, M.L., G. De La Rosa, J.A. Hernandez-Viezcas, J.R. Peralta-Videa and J.L. Gardea-Torresdey. 2010. X-ray absorption spectroscopy (XAS) corroboration of the uptake and storage of CeO_2 nanoparticles and assessment of their differential toxicity in four edible plant species. Journal of Agricultural Food Chemistry 58: 3689–3693.

Lu, C.M., C.Y. Zhang, J.Q. Wen and G.R. Wu. 2002. Effects of nano material on germination and growth of soybean. Soybean Science 21: 168–171.

Moraru, C., C. Panchapakesan, Q. Huang, P. Takhistov and S. Liu. 2003. Nanotechnology: A new frontier in food science. Food Technology 57: 24–29.

Parsons, J.G., M.L. Lopez, C.M. Gonzalez, J.R. Peralta-Videa and J.T. Gardea-Torresdey. 2010. Toxicity and biotransformation of uncoated and coated nickel hydroxide nanoparticles on mesquite plants. Environmental Toxicology and Chemistry 29: 1146–1154.

Pradhan, S., P. Patra, S. Das, S. Chandra, S. Mitra, K.K. Dey, S. Akbar, P. Palit and A. Goswami. 2013. Photochemical modulation of biosafe manganese nanoparticles on Vigna radiata: A detailed molecular, biochemical, and biophysical study. Environmental Toxicology and Chemistry 47: 13122–13131.

Pradhan, S., I. Roy, G. Lodh, P. Patra, S.R. Choudhury, A. Samanta and A. Goswami. 2011. Entomotoxicity and biosafety assessment of PEGylated acephate nanoparticles: A biologically safe alternative to neurotoxic pesticides. Journal of Environmental Science and Health, Part B 48: 559–569.

Racuciu, M. and D.E. Creanga. 2009. Biocompatible magnetic fluid nanoparticles internalized in vegetal tissue. Romanian Journal of Physics 54: 115–124.

Salata, O. 2004. Applications of nanoparticles in biology and medicine. Journal of Nanobiotechnology 2.

Shen, C.X., Q.F. Zhang, J. Li, F.C. Bi and N. Yao. 2010. Induction of programmed cell death in *Arabidopsis* and rice by single-wall carbon nanotubes. American Journal of Botany 97: 1–8.

Sheykhbaglou, R., M. Sedghi, M.T. Shishevan and R.S. Sharifi. 2010. Effects of nano-iron oxide particles on agronomic traits of soybean. Notulae Scientia Biologicae 2: 112–113.

Stampoulis, D., S.K. Sinha and J.C. White. 2009. Assay-dependent phytotoxicity of nanoparticles to plants. Environmental Science and Technology 43: 9473–9479.

Tan, X.M., C. Lin and B. Fugetsu. 2009. Studies on toxicity of multiwalled carbon nanotubes on suspension rice cells. Carbon 47: 3479–3487.

Tan, X.M. and B. Fugetsu. 2007. Multi-walled carbon nanotubes interact with cultured rice cells: Evidence of a self-defense response. Journal of Biomedical Nanotechnology 3: 285–288.

Taiz, L. and E. Zeiger. 1998. Plant Physiology. 2nd ed., Sinauer: Sunderland, MA.

Wang, H., X. Kou, Z. Pei, J.Q. Xiao, X. Shan and B. Xing. 2013. Physiological effects of magnetite (Fe_3O_4) nanoparticles on perennial ryegrass (*Lolium perenne* L.) and pumpkin (*Cucurbita mixta*) plants. Nanotoxicology 5: 32–40.

Yang, F., F.S. Hong, W.J. You, C. Liu, F.Q. Gao, C. Wu and P. Yang. 2006. Influences of nano-anatase TiO_2 on the nitrogen metabolism of growing spinach. Biological Trace Element Research 110: 179–190.

Yang, F., C. Liu, F. Gao, M. Su, X. Wu, L. Zheng, F. Hong and P. Yang. 2007. The improvement of spinach growth by nano-anatase TiO_2 treatment is related to nitrogen photoreduction. Biological Trace Element Research 119: 77–88.

Zhao, J. and V. Castranova. 2011. Toxicology of nanomaterials used in nanomedicine. Journal of Toxicology and Environmental Health, Part B. Critical Reviews 14: 593–632.

Zheng, L., F. Hong, S. Lu and C. Liu. 2005. Effect of nano-TiO_2 on strength of naturally aged seeds and growth of spinach. Biological Trace Element Research 104: 83–91.

CHAPTER 9

Nano Zinc Micronutrient

M. Yuvaraj[1] and *K.S. Subramanian*[2,*]

1. Introduction

The importance of agriculture to all human societies is characterized more than ever with the increasing world population. The first and most important need of every human is access to food, and the food supply for humans is directly or indirectly associated with agriculture. Globally, about two billion people are affected by Zn deficiency, where people are exposed to cereal based diets with deficient level of Zn (Welch and Graham 2004). Most children suffering from Zn deficiency have stunted linear growth. Recommended daily intakes range between 3 and 16 mg Zn day[-1] depending on age, gender, type of diet and other factors. It has been estimated that around 33% of the world's human population has diets deficient in Zn, but this ranges between 4 and 73% in different countries (Hotz and Brown 2004). The Food and Agriculture Organization of the United Nations (FAO 2002) estimates that 50% of world's soils growing cereal grains are zinc deficient. It also estimates that agricultural production must increase by 70% by 2050 in order to feed over 9 billion people worldwide. Zinc deficiency, especially in infants and young children under five years of age, has received global attention. Zinc deficiency is the fifth leading cause of death and disease in the developing world. According to the World Health Organization (WHO 2002) about 800,000 people die annually due to zinc deficiency, of which 450,000 are children under the age of five. Globally, around 2 billion people are affected by zinc deficiency.

The issue of micronutrient deficiency is related to food security (Meenakshi et al. 2010). Micronutrient deficiencies in human being as well as crop plants are difficult to diagnose and accordingly the problem is termed as 'hidden hunger' (Stein et al. 2008). This hidden hunger may cause nearly 40 percent reductions in crop productivity and it is also estimated that it affects more than a half of the global

[1] Adhiparasakthi Agricultural College, Kalavai, Vellore, Tamil Nadu, India.
[2] Department of Nano Science and Technology, Tamil Nadu Agricultural University, Coimbatore, Tamil Nadu, India.
* Corresponding author

population. Micronutrient deficiency in general refers to Fe, Zn, Se, I, Cu, Ca and Mg (Zhao and Mcgrath 2009); among them, zinc deficiency is most wide spread, followed by Iron. World Health Organization reported that Zn deficiency stands as the fifth greatest risk factor for causing diseases among children in developing countries. Based on analysis of diet composition and nutritional needs, it has been estimated that 49% of the world's population (equivalent to 3 billion) are at risk of suffering from Zn deficiency. Until recent times, soil fertilization was the only way to meet the mineral requirement of crops plants. However, several problems exist, such as the need for large quantities of fertilizer, fixation in soil and slow uptake by plants.

In order to overcome Zn disorder, several strategies are being employed, including supplementation, fortification, diversification and biofortification. Among these strategies, in recent years the application of nano scale particles of Zn is preferred for enhancing the agronomic effectiveness of Zn fertilizers. Nanofertilizers are an important tool in agriculture for improving crop growth, yield and quality parameters with increased nutrient use efficiency and reduced wastage of fertilizers and cost of cultivation. Nanofertilizers are very effective for precise nutrient management in precision agriculture, matching the crop growth stage with the correct nutrients and providing nutrients throughout the crop growth period. Nano-fertilizers increase crop growth up to optimum concentrations, however, increase in concentration may inhibit the crop growth due to the toxicity of the nutrients. Nano-fertilizers provide more surface area for different metabolic reactions in the plant which increase the rate of photosynthesis and produce more dry matter and yield of the crop. It also protects plant from different biotic and abiotic stress.

Nanotechnology deals with atom-by-atom manipulation and the processes and products evolved are quite precise. Despite the fact that the nanotechnology is highly exploited in the field of energy, environment and health, the research in agricultural sciences has only scratched the surface. However, the potentials of nanotechnology in agricultural sciences had been reviewed (Subramanian and Tarafdar 2011). Among the applications, nanofertilizer technology is very innovative and known to exhibit economic advantages if the products evolved are economically feasible and socially sustainable. These customized fertilizers are reported to reduce nutrient loss caused by leaching, emissions, and long-term incorporation by soil microorganisms (DeRosa et al. 2010).

Fertilizers encapsulated in nanoparticles increase the uptake of nutrients. In the next generation of nanofertilizers, the release of the fertilizer can be triggered by environmental conditions or simply released over time. Slow, controlled-release fertilizers have the potential to increase the efficiency of nutrient uptake. Nanofertilizers that utilize natural materials for coating and cementing granules of soluble fertilizer have the advantage of being less expensive to produce than those fertilizers that rely upon manufactured coating materials. Slow, controlled-release fertilizers may also improve soil by decreasing the toxic effects associated with over-application of fertilizer.

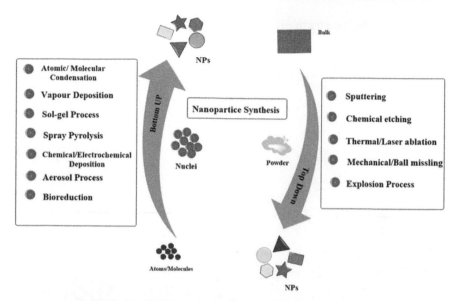

Fig. 1. Top-down and bottom up methods of synthesizing nano-particles (NPs).

One of the most commonly used mesoporous materials used for the synthesis of nano-fertilizers, zeolites, have been used as a slow release nutrient delivery system. Zeolites is a naturally occurring mineral group consisting of about 50 mineral types that draws attention as a good growing medium substrate for a long period due to its good physical and chemical characteristics (Markovich et al. 1995). They have a rigid three-dimensional crystal structure with voids and channels of molecular size and high cation exchange capacity (CEC) arising from substitution of Al for Si in the silicon oxide tetrahedral units that constitute the mineral structure (Pickering et al. 2002). The features of zeolites are: (1) high nutrient absorption capacity (2) retains water and nutrients (3) slowly releases nutrient as slow release fertilizers.

In Tamil Nadu Agricultural University, Coimbatore, several nanozeolite-based fertilizer formulations carrying N (Mohanraj 2013, Manikandan and Subramanian 2014), P and K (Sharmila Rahale 2010), S (Selva Preetha et al. 2012, Thirunavukkarasu and Subramanian 2014) and Zn (Subramanian et al. 2012), Zn (Yuvaraj and Subramanian 2014, 2018) have been synthesized, characterized and tested in various crops. These formulations are slow release nutrient carriers with higher use efficiency.

Nanotechnology has progressively moved away from the experimental into the practical areas, for instance, the development of slow or controlled release fertilizers on the basis of nanotechnology has become critically important for promoting the development of environmentally friendly and sustainable agriculture. Indeed, nanotechnology has provided the feasibility of exploiting nanoscale or nanostructured materials as fertilizer carriers or controlled release vectors for building of so-called "smart fertilizer" as new facilities to enhance nutrient use

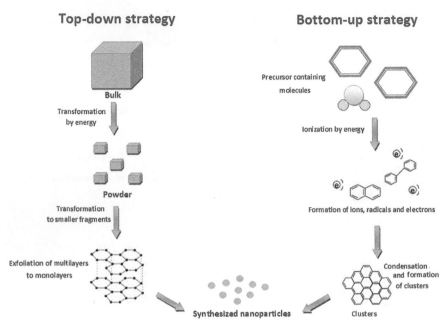

Fig. 2. Schematic illustration showing synthesis of nano-particles (NPs) using top-down and bottom up approaches.

efficiency and reduce costs of environmental protection. Encapsulation of fertilizers within a nanoparticle is one of these new facilities which are done in three ways: (a) The nutrient can be encapsulated inside nanoporous materials, (b) coated with thin polymer film and (c) delivered as particle of nanoscales dimensions. In addition, nanofertilizers will combine nanodevices in order to synchronize the release of Zn fertilizer and the uptake by crops, so preventing undesirable nutrient losses to soil, water and air via direct internalization by crops, and avoiding the interaction of nutrients with soil, microorganisms, water and air.

In recent years, nanotechnology has extended its relevance in plant science and agriculture. Advancement in nanotechnology has improved ways for large-scale production of nanoparticles of physiologically important metals, which are now used to improve fertilizer formulations for increased uptake in plant cells and minimized nutrient loss. Nanoparticles have high surface area, sorption capacity, and controlled-release kinetics to targeted sites, making them a "smart delivery system". Nanostructured fertilizers can increase the nutrient use efficiency through mechanisms such as targeted delivery, slow or controlled release. They could release their active ingredients with precision in response to environmental triggers and biological demands. In recent lab scale investigations, it has been reported that nano-fertilizers can improve crop productivity by enhancing the rate of seed germination, seedling growth, photosynthetic activity, nitrogen metabolism, and carbohydrate and protein synthesis. However, being an infant technology, the ethical and safety issues surrounding the use of nanoparticles in plant productivity are limitless and must be very carefully evaluated before adapting the use of the so-called nano-fertilizers in agricultural fields.

In this regard, inclusion of zeolites is known to provide extensive surface area and is capable of regulating the adsorption and desorption of nutrients that eventually increase the crop yields (Ramesh et al. 2010). Utilization of zeolites in agriculture is possible because of their special cation exchange properties, molecular sieving and adsorption (Mumpton 1999). Blending fertilizers with zeolites can produce the same yield from less fertilizer applied because of the reduction of volatilization and leaching losses. Zeolites can hold nutrients in the root zone that lead to improved use efficiencies of plant nutrients. The large honeycomb crystal structure provides a huge storage space. Cationic plant nutrient ions, e.g., Potassium (K^+) and zinc (Zn^{2+}), and water molecules are also stored in the zeolite crystal and are readily available to the plant.

The adsorptive surface area of zeolites can further be improved by ball milling process. Ball milling is top–down approach, used to reduce the size of the zeolite in order to increase the adsorptive sites, where cations like Zn^{2+} can adsorb and desorb slowly, matching with the crop demand. Nano-zeolite may show great promise as anion and cation carriers to control Zn release. Nano-zeolite are negatively changed and they are capable of adsorbing Zn^{2+} ions that facilitate slow, steady and regulated release of nutrients. This process will result in nano-fertilizer formulations that assist in the regulated release of nutrients and improve Zn use efficiency while preventing environmental hazards (Chirenje and Liping 2005). Nano-zeolites loaded with zinc sulphate ($ZnSO_4$) increased the Zn use efficiency of certain vegetables, such as lettuces. This study hypothesized that nano-zeolites possess extensive surface area that can adsorb Zn^{2+} ions abundantly, thereby regulating the release of Zn to the soil solution. The nano-Zn fertilizer developed using nano-zeolite as a carrier was characterized using a standard set of protocols. Thereafter, the same material was assessed for its Zn release and fractionation patterns.

2. Nano-Fertilizers for Balanced Crop Nutrition

In India, fertilizers, along with quality seed and irrigation, are mainly responsible for the enhanced food grain production from 1960 (55 mt) to 2011 (254 mt), coinciding with the vast increase in fertilizer consumptions from 0.5 mt to 23 mt, respectively. It has been conclusively demonstrated that fertilizers contribute to the tune of 35–40 per cent of the productivity of any crop. Considering their importance, the Government of India is heavily subsidizing the cost of fertilizers, particularly urea. This has resulted in imbalanced fertilization and nutrient deficiency occurrence in some areas, as well as nitrate pollution of ground waters due to excessive nitrogen application. In the past few decades, use efficiencies of N, P and K fertilizers have remained constant at 30–35 per cent, 18–20 per cent and 35–40 per cent, respectively, leaving a major portion of added fertilizers to accumulate in the soil or enter into aquatic system causing eutrophication. In order to address issues of low fertilizer use efficiency, imbalanced fertilization, multi-nutrient deficiencies and decline of soil organic matter, it is important to evolve a nano-based fertilizer formulation with multiple functions. Significant increase in yields has been observed due to foliar application of nano particles as fertilizer (Tarafdar et al. 2012). It was shown that 640 mg/ha foliar application (40 ppm concentration) of nanophosphorous gave 80 kg/ha P equivalent yield of clusterbean and pearl millet in an arid environment. Currently, research to develop nano-composites to supply all the required essential nutrients in suitable proportion through smart delivery system is underway. Preliminary results suggest that balanced fertilization may be achieved through nanotechnology (Tarafdar et al. 2012). Indeed, the metabolic assimilation within the plant biomass of the metals, e.g., micronutrients, applied as nano-formulations through soil-borne and foliar application or otherwise, needs to be ascertained. Further, the nano-composites being contemplated to supply all the

Table 1. Unique properties and advantages of nano-fertilizers in comparison to conventional fertilizers.

Properties	Nano Fertilizer Enabled Technologies	Conventional Fertilizer
Solubility and availability	Nano-sized formulation improves solubility and reduced soil absorption, fixation and higher the bioavailability	Less bioavailable due to large particle size
Nutrient uptake efficiency	Nanostructured formulation might increase fertilizer efficiency and uptake ratio of the soil nutrients in crop production and save fertilizer resource	Bulk composite is not available for roots and decreases efficiency
Controlled release modes	Both release rate and release pattern of nutrients controlled by encapsulation in envelope form of semipermeable membranes coated by resin-polymer, waxes and sulphur	Excess release of fertilizer may produce toxicity and destroy ecological balance of soil
Effective duration of nutrients release	Nanostructured formulation can extend effective duration of nutrient supply of fertilizer in to soil	Used by the plant at the time of delivery the rest is converted into insoluble salts in the soil
Loss rate of fertilizer nutrients	Nanostructured formulation can reduce loss rate of fertilizer nutrients	High loss rate by leaching, run off and drift

Cui et al. 2010

nutrients in the right proportions through the "smart" delivery systems also need to be examined closely. Currently, the nitrogen use efficiency is low, as a loss of 50–70 per cent of the nitrogen supplied in conventional fertilizers is experienced. New nutrient delivery systems that exploit the porous nano scale parts of plants could reduce nitrogen loss by increasing plant uptake. Fertilizers encapsulated in nanoparticles will increase the uptake of nutrients (Tarafdar et al. 2012). In the next generation of nano fertilizers, the release of the nutrients can be triggered by an environmental condition or simply released at a desired specific time.

Table 2. Crop responses to various nano-fertilizers.

S. No.	Nano Fertilizer	Crop Used	References
1	Nano zinc oxide	Ground Nut	Prasad et al. 2012
2	Zinc nano fertilizer	Pearl Millet	Tarafdar et al. 2014
3	Nano K-fertilizer	Rice	Sirisena et al. 2013
4	Nano calcium	Tomato	Tantawy et al. 2014
5	Nano zinc sulphide	Sunflower	Meena 2015
6	Zinc oxide nano particle	Strawberry	Jayvanth et al. 2017
7	Nano carbon	Rice	Mei-yan 2010
8	Nano rock phosphate	Maize	Adhikari et al. 2014
9	Nano sulphur	Ground Nut	Tirunavukarasau et al. 2014
10	Nano zinc	Maize	Sharmila Rahale et al. 2012
11	Zeo zinc	Rice	Yuvaraj et al. 2014
12	Manganese core shell	Rice	Yuvaraj et al. 2014
13	Carban sphere	Rice	Yuvaraj et al. 2014
14	Zinc oxide	Rice	Yuvaraj et al. 2014

3. Smart Delivery Systems

Nanoscale devices are envisioned that would have the capability to detect and treat diseases, nutrient deficiencies or any other maladies in crops long before symptoms are visually exhibited. "Smart Delivery Systems" for agriculture can possess timely controlled, spatially targeted, self-regulated, remotely regulated, preprogrammed, or multi-functional characteristics in order to avoid biological barriers to achieve the target. Smart delivery systems can monitor the effects of delivery of nutrients or bioactive molecules or any pesticide molecules. This is widely used in health sciences, wherein nanoparticles are exploited in order to deliver required quantities of medicine to the place of need in human systems. In the smart delivery system, a small sealed package carries the drug which opens up only when the desirable location or infection site of the human or animal system is reached. This would allow judicious use of antibiotics that otherwise would be possible. Nanotechnology will revolutionize agriculture and food industry through innovation and new techniques, such as precision farming techniques, enhancing the ability of plants to absorb nutrients, and more efficient and targeted use of inputs.

4. Application of Nano Zinc Oxide in Agriculture

In recent years the application of nano scale particles of Zn is being preferred for enhancing the nutrient uptake by plants. Particle size may affect agronomic effectiveness of Zn fertilizers. Decreased particle size results in increased number of particles per unit weight of applied Zn and also increases the specific surface area of a fertilizer, which will increase the dissolution rate of fertilizers with low solubility in water, such as Zn oxide (ZnO) (Mortvedt 1992). Granular Zn sulphate ($ZnSO_4$) (1.4 to 2 mm) was somewhat less effective than fine $ZnSO_4$ (0.8 to 1.2 mm), and granular ZnO was completely ineffective (Allen and Andrews 1977). Gradual increase in Zn uptake could be observed with decreasing granule size. Since granules of 1.5 mm weigh less than granules of 2.0 or 2.5 mm, smaller

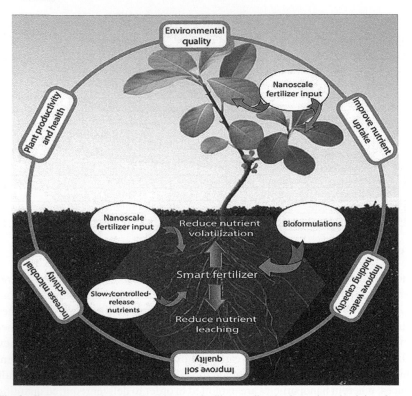

Fig. 3. General mechanisms involved in nano-fertilizer mediated enhanced productivity of crops.

granules were used for the same weight, resulting in a better distribution of Zn and the higher surface area of contact of Zn fertilizer resulted in better Zn uptake. Therefore, ample work has been done and emphasis was made on the particle size in order to increase the efficiency of the fertilizers for better uptake and higher yields. The schematic illustration showing the role of nano-fertilizers in promoting productivity of crops is detailed in Fig. 9.3.

Effectiveness of ZnO on seed germination, seedling vigour, plant growth, flowering, chlorophyll content, pod yield and root growth of peanut was studied. Peanut seeds were treated with different concentrations of nano ZnO along with chelated Zn Sulphate ($ZnSO_4$). Treatment of nano scale ZnO (25 nm mean particle size) at 1000 ppm concentration promoted both seed germination and seedling vigour and, in turn, seedlings showed early establishment in soil, higher leaf chlorophyll content, which manifested itself in early flowering. These particles proved effective in increasing stem and root growth. Pod yield per plant was 34 per cent higher compared to chelated bulk $ZnSO_4$.

Spraying optimum concentrations of nano-ZnO, nano-FeO and nano-ZnCuFeoxide in suspension form on hydroponically grown test units of Mung (*Vigna radiata*) seedling and examining the effect on the shoot growth of seedlings revealed that the seedlings displayed good growth in comparison to the control, demonstrating a positive effect of the nanoparticle treatment. The best performance was observed for nano Zn/CuFe-Oxide, followed by nano FeO and nano ZnO. Increased absorption of nanoparticles by plant leaves was also detected by inductive coupled plasma/atomic emission spectroscopy.

These ZnO NPs were used during the seed germination and root growth of *Cicer arietinum*. The effect of ZnO NPs has been observed on the seed germination and root growth of Cicer arietinum seeds. ZnO nanoparticles (NPs) have been synthesised by hydrothermal method. This hydrothermally synthesised product has been characterised by powder X-ray diffraction and field emission scanning electron microscopy (FE-SEM) for the study of crystal structure and morphology/size. FESEM image revealed that ZnO NPs are spherical in shape with a diameter of 20–30 nm.

The effect of nano ZnO particles on the growth of plant seedlings of mung (*Vigna radiate*) and gram (*Cicer arietinum*) was studied by conducting experiments in agar media. Various concentrations of nano ZnO particles in suspension form were introduced to the agar media and their effect on the root and shoot growth of the seedlings was examined. The main experimental approach, using correlative light and scanning electron microscopy, provided evidence of adsorption of nanoparticles on the root surface. Absorption of nanoparticles by seedlings root was also detected by inductive coupled plasma/atomic emission spectroscopy (ICP-AES). It was found that at certain optimum concentrations, the seedlings displayed good growth in comparison to the control and, beyond that, retardation in growth was observed due to toxicity.

Different concentrations (0, 10, 20, 30 and 40 g ml/1) of nano ZnO particles were used for the treatment in onion seeds to study the effect on cell division, seed germination and early seedling growth. Investigators observed decreased Mitotic Index (MI) and increase in chromosomal abnormalities in higher treatments of zinc oxide nanoparticles. Seed germination increased in lower concentrations, but showed decrease in values at higher concentrations. Germination indices showed increased values in lower concentrations; however, these decreased significantly at higher concentrations.

The effects of bulk and nano dimensional zinc oxide particles on germination, growth and biochemical parameters of cabbage, cauliflower and tomato crops were studied. It was observed that bulk ZnO particles are phytotoxic and adversely affect germination, seedling growth and biochemical parameters of all the test crops. They decreased germination and caused stunting of stems. Unlike the bulk, nanodimensional zinc oxide particles enhanced germination, seedling growth, pigments, sugar and protein contents, along with increased activities of antioxidant enzymes in all the three test crops. Zinc oxide NPs invariably increase pigments, protein and sugar contents and nitrate reductase activities in cabbage (*Brassica oleracea* var. Capitata). Cauliflower, in response to higher concentration of ZnO NPs (9.0 µM), maintained the germination, seedling growth and sugar at the level of the control, while SOD and CAT to the level ZnO treated seedlings. ZnO NPs induced activities of antioxidant enzymes, viz. SOD, CAT, APX and POD. The activities of antioxidant enzymes in ZnO NP treated seedlings were lower as compared to those treated with bulk ZnO. Nanoparticles buttressed the metabolic processes of the three test crops. The order of sensitivity of the crops to ZnO and ZnO NPs was cauliflower < cabbage. ZnO NPs appeared to mitigate the phytotoxic effects of bulk.

5. Importance of Nano-Zinc in Drought Stress Management

In order to study the effect of nano zinc oxide on germination parameters of soybean seeds under drought stress conditions, a group of researchers conducted an experiment by inducing stress through application of poly ethylene glycol (PEG). Treatments were drought stress (0, –0.5, –1 MPa) and concentrations of nano zinc oxide (0, 0.5, 1 g/lit). Results showed the effect of different concentrations of PEG and nano zinc oxide on germination. Nano zinc oxide increased germination percentage rate in comparison to the control. Length and fresh weight of radical was greater in stressed seedlings.

Application of nano zinc oxide and stress occurrence caused a decrease in seed residual fresh and dry weight, which indicates that treatments were effective for using of seed reservoirs to seedling growth. Mesquite (Prosopis juliflora-velutina) seedlings were grown for 15 days in hydroponics with ZnO NPs (10 nm) at concentrations varying from 500 to 4000 mg/l. Zn concentrations in roots, stems and leaves were determined by inductively coupled plasma optical emission spectroscopy (ICP-OES). Plant stress was examined by the specific activity of catalyse (CAT) and ascorbate peroxidase (APOX); while the biotransformation of ZnO NPs and Zn distribution in tissues was determined by X-ray absorption spectroscopy (XAS) and micro X-ray fluorescence (XRF), respectively. ICP-OES results showed Zn concentrations of 2102 ± 87, 1135 ± 56, and 628 ± 130 mg/kg dry weight in roots, stems and leaves, and tissues, respectively, at 2000 mg/l ZnO NPs. Stress tests showed that ZnO NPs increased CAT in roots, stems and leaves, while APOX increased only in stems and leaves. XANES spectra demonstrated that ZnO NPs were not present in mesquite tissues while Zn was found as Zn^{2+}, resembling the spectra of Zn. The XRF analysis confirmed the presence of Zn in the vascular system of roots and leaves in ZnO NP treated plants.

6. Advantages of Nano-Fertilizers

Nano fertilizers have an important role in crop production, and several research studies revealed that nano fertilizers enhanced growth, yield and quality parameters of the crop, which results in a better yield and quality of food product for human and animal consumption. This translates into an improvement to three major areas of production. Several research studies revealed that application of nano-fertilizers significantly increases crop yield in comparison to the control or no application of nano-fertilizer, mainly because the increased growth of plant parts and metabolic processes such as photosynthesis leads to higher photosynthates accumulation and translocation to the economic parts of the plant. Foliar application of nano particles as fertilizer significantly increases yield of the crop. Nano fertilizers provide more surface area and more availability of nutrient to the crop plant which help to increase these quality parameters of the plant (such as protein, oil content and sugar content) by enhancing the rate of reaction or synthesis process in the plant system. Application of zinc and iron on the plant increase total carbohydrate, starch, IAA, chlorophyll and protein content in the grain. NanoFe$_2$O$_3$ increases photosynthesis and growth of the peanut plant.

7. Conclusion

Application of different nano-fertilizers has a greater role in enhancing crop production, and will reduce the cost of fertilizer for crop production and also minimize the pollution hazard. The application of nano-fertilizers in agriculture should be given more attention. Fertilizer nutrient use efficiency in crop production can be enhanced with effective use of nano-fertilizers. Nano fertilizers improve crop growth and yield up to optimum applied doses and concentrations, but they also have inhibitory effect on crop plants if concentration is more than the optimum, resulting in reduced growth and yield of the crop.

Encapsulation of fertilizers within a nanoparticle is one of these new facilities which are done in three ways (a) the nutrient can be encapsulated inside nano porous materials, (b) coated with thin polymer film, (c) delivered as particle or emulsions of nano scales dimensions. In addition, nano fertilizers will combine nano devices in order to synchronize the release of fertilizer-N and -P with their uptake by crops, thereby preventing undesirable nutrient losses to soil, water and air via direct internalization by crops, and avoiding the interaction of nutrients with soil, microorganisms, water and air.

World population is increasing exponentially, so it is a great challenge for the agriculture sector to feed the growing population with nutritious food. There are several biotic and abiotic constraints which limit agricultural productivity. Nutritional stress is the third most important abiotic stress, next to drought and salinity stress, which not only limits the growth and productivity of plants but also has an impact on human health. Plants play a vital role in human nutrition by providing all essential nutrients required for human health. However, they are not in a position to meet the daily requirement of mineral nutrients due to the fact that most of these essential nutrients or elements, especially micronutrients, are not present in the required concentrations. Most of our food crops contain lower amount of Zn in their edible parts. Thus biofortification of food grains with micronutrients is essential to overcome nutritional disorders in humans. Globally, the soils are Zn deficient and plants are not in a position to take up, translocate and accumulate enough Zn in edible parts to meet the human nutrition requirement. Hence, in order to correct this problem, the prior information on physiological, biochemical and molecular mechanisms that contribute to the uptake, transport and accumulation of Zn in plants is crucial. Based on this information, it will be easy for developing strategies to manipulate the crops for their nutritional status.

Biofortification of food crops is essential in order to increase the zinc content in edible parts. Direct application of Zn fertilizers to the crop at the appropriate stage can significantly boost the Zn content in the plants' edible parts. Since zinc is a heavy metal, excess application leads to accumulation in soil at toxic levels, causing injury to the plants. In this regard, careful strategy should be adopted to minimize this toxic effect. Recently, the application of nano zinc is preferred for enhancing the zinc content in the edible parts. Because the nano Zn fertilizers have high surface area to volume ratio, they are only required in very small quantities and promote better absorption and translocation, thereby reducing the zinc toxicity. With this concern, major emphasis was given in the present study to efficacy on yield and Zn content in the edible parts of some common vegetable crops by using nano ZnO, compared with granular $ZnSO_4$, which is used as a common Zn source.

References

Allen, E. and R. Andrews. 1997. Space age soil mix uses centuries-old zeolites. Golf Course Manage. 61(12): 24–28.

Chirenje, T., Q. Ma and Liping Lu. 2005. Retention of Cd, Cu, Pb and Zn by Ash, Lime and Fume Dust. Springer, pp. 301–315.

DeRosa, M.C., C. Monreal, M. Schnitzer, R. Walsh and Y. Sultan. 2010. Nanotechnology in fertilizers. Nature Nanotechnol. 32(5): 1234–1237.

FAO. 2002. Rice Information, Volume 3. Food and Agriculture Organization of United Nations, Rome. Available from: http://www.fao.org/docrep/005/y4347e/y4347e00.htm.

Hotz, C. and K.H. Brown. 2004. Assessment of the risk of zinc deficiency in population and options for its control. Food and Nutr Bull. 25: 94–203.

Manikandan, A. and K.S. Subramanian. 2014. Fabrication and characterisation of nanoporous zeolite-based N fertilizer. African Journal of Agriculture Research 9(2): 276–284.

Markovich, A., A. Takac, Z. Illin, T. Ito and F. Tognoni. 1995. Enriched zeolites as substrate component in the production of paper and tomato seedling. Acta Horticulturae 39(6): 321–328.

Mohanraj, J. 2013. Effect of nano-zeolite on nitrogen dynamics and greenhouse gas emission in rice soil eco system. M.Tech, Thesis, Tamil Nadu Agricultural University, Coimbatore.

Mortvedt, J.J. and P.M. Giordano. 1969. Extractability of zinc granulated with macronutrient fertilizers in relation to its agronomic effectiveness. J. Agr. Food Chem. 17: 1272–1275.

Mumpton, F.A. 1999. La roca magica: Use of natural zeolite in agriculture and industry. Proc. Nat. Acad. Sci. 96: 3463–3470.

Pickering, H.W., N.W. Menzies and M.N. Hunter. 2002. Zeolite rock phosphate-a novel slow release phosphorus fertiliser for potted plant production. Scientia Horticulturae 9(4): 333–343.

Ramesh, K., A.K. Biswas, J. Somasundaram and A. Subbarao. 2010. Nanoporous zeolites in farming: Current status and issues ahead. Curr. Sci. 99:760–764.

Selva Preetha, P. 2012. Nano-fertilizer formulation to achieve balanced nutrition in greengram (*Vigna radiata*). M.Sc., Thesis, Tamil Nadu Agricultural University, Coimbatore.

Sharmila Rahale, C. 2010. Nutrient release pattern of nano-fertilizer formulations. Ph.D., Thesis, Tamil Nadu Agricultural University, Coimbatore.

Subramanian, K.S. and C. Sharmila Rahale. 2012. Ball milled nanosized zeolite loaded with zinc sulphate: A putative slow release Zn fertilizer. Interna. J. Indian Hort. 1(1): 33–40.

Subramanian, K.S. and J.C. Tarafdar. 2011. Prospects of nanotechnology in Indian farming. Indian J. Agr. Sci. 81: 887–893.

Tarafdar, J.C. 2012. Perspectives of nanotechnological applications for crop production.

Thirunavukkarasu. M. and K.S. Subramanian. 2014. Surface modified nano-zeolite used as carrier for slow release of sulphur. Journal of Applied and Natural Science 6(1): 19–26.

Welch, R.M. and R.D. Graham. 2004. Breeding for micronutrients in staple food crops from a human nutrition perspective. Journal of Experimental Botany 55: 353–364.

WHO. 2002. The world health report reducing risks, promoting healthy life. World Health Organization, Geneva Switzerland.

Yuvaraj, M. and K.S. Subramanian. 2014. Controlled-release fertilizer of zinc encapsulated by a manganese hollow core shell. Soil Science and Plant Nutrition. doi:10.1080/00380768. 2014.979327.

Yuvaraj, M. and K.S. Subramanian. 2018. Development of slow release Zn fertilizer using nano-zeolite as carrier. Journal of Plant Nutrition 41(3): 311–320. doi:10.1080/01904167.2017.1381729.

CHAPTER 10

Thymol Based Nanoemulsions

Current and Future Prospects in Disease Protection and Plant Growth

Sarita Kumari,[1] Kumara Swamy R.V.,[1]
Ram Chandra Choudhary,[1] Savita Budhwar,[2] Ajay Pal,[3]
Ramesh Raliya,[4] Pratim Biswas[4] and Vinod Saharan[1,]*

1. Introduction

Essential oils contain a complex mixture of non-volatile and volatile compounds, produced by aromatic plants as secondary metabolites, with a wide-spectrum of biological activities (Esmaeili and Asgari 2015). The fact that essential oils are considered as "natural" components makes them highly desirable for use in many food products, since there is growing consumer demand for natural rather than synthetic additives. The major constituents in commercial essential oils can be classified into three classes: Phenols, terpenes and aldehyde (Chang et al. 2012). Thymol (2-isopropyl-5-methylphenol) (Fig. 1) is a major essential oil phenolic component (EOCs) of the thyme oil extracted from the herb *Thymus vulgaris* and is classified as generally recognized as safe (GRAS) by the U.S. Food and Drug Administration. It is also known as "hydroxy cymene" because it is biosynthesized

[1] Department of Molecular Biology and Biotechnology, Rajasthan College of Agriculture, Maharana Pratap University of Agriculture and Technology, Udaipur, Rajasthan 313001, India.
[2] School of Interdisciplinary and Applied Life Sciences, Central University of Haryana, Mahendergarh, Haryana 123031, India.
[3] Department of Biochemistry, College of Basic Sciences and Humanities, Chaudhary Charan Singh Haryana Agricultural University, Hisar, Haryana 125004, India.
[4] Department of Energy, Environmental and Chemical Engineering, Washington University in St. Louis, St. Louis, MO, 63130, USA.
* Corresponding author: vinodsaharan@gmail.com

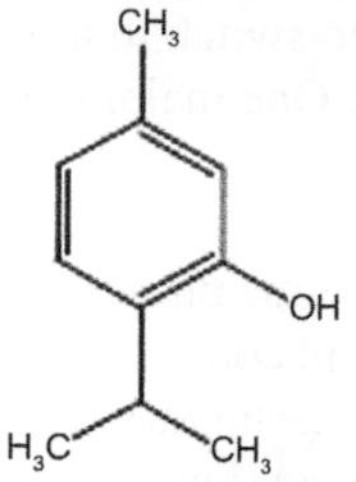

Fig. 1. Chemical structure of thymol.

by the aromatization of γ-terpinene to p-cymene followed by hydroxylation of p-cymene. The antimicrobial efficacy of thymol has been known for decades (Poulose and Croteau 1978).

Thymol exhibits strong antimicrobial activity against a wide range of bacteria and fungi (Chang et al. 2012). It contains a phenolic hydroxyl group that makes it a promising antimicrobial by allowing it to penetrate into and disturb the structure of microbial cell membranes, thereby leading to leakage of ions and other cell contents (Di Pasqua et al. 2007). Thymol has been reported to exhibit antifungal effect by disrupting ergosterol biosynthesis and cell membrane integrity (Ahmad et al. 2011). Its use has been broadly reported in the fields of medicine (Silva et al. 2011), food (Evans and Martin 2000, Lambert et al. 2001, Sacchetti et al. 2005, Oussalah et al. 2006, Shapira and Mimran 2007) and agriculture (Glenn et al. 2010, Gill et al. 2016, Kumari et al. 2018). In recent times, thymol has been exploited as a plant growth regulator in crops and can be exploited for agrochemical delivery through nanotechnology. However, there are some limitations concerning the use of thymol in aqueous medium, such as instability, insolubility in aqueous medium and its highly volatile nature (Shoji and Nakashima 2004). To explore the potential of thymol for crop plants it is important to study its physico-chemical dynamics in order to improve its characters for easy and efficient use. Therefore, it is imperative to find some methods to increase the stability, solubility and controlled release of thymol in aqueous systems (Chen et al. 2015). In this regard, the nanoemulsion technology seems to be a potential strategy to enhance its solubility, stability, efficacy and protection (Donsi et al. 2011). Nanoemulsions have been considered efficient delivery systems for hydrophobic compounds by dispersing the lipid phase as a colloidal dispersion. Thymol nanoemulsion has good kinetic stability and retains the antimicrobial and antioxidant properties of thymol. Nanoemulsions are a commercially valuable delivery system because they have the unique characteristics of small size and high surface area, optical clarity, and reduced rate of gravitational separation and flocculation. Nanoemulsions are commonly stabilized by amphiphilic surfactants which are absorbed between water and oil phases. Surfactants perform two functions in nanoemulsions: (1) reduce interfacial tension between oil and water phase (2) decrease the rate of coalescence of oil droplets by forming a physical, steric and/or electric barrier around them. Synthetic surfactants are widely used in emulsion but the potential toxicity of synthetic surfactant limits their usage in food system. It is therefore desirable to switch to a greener alternative surfactant with a reduced environmental burden. One major constraint faced by researchers post nanoemulsion production is the process of Oswald ripening, where the mean size of the particle increases with time due to diffusion of oil molecules from small to large droplets through the continuous phase. A possible way of overcoming this instability to increase the surfactant concentration by altering the oil-to-surfactant ratio. This chapter reviews various techniques for thymol nanoemulsion preparation and applications of thymol and thymol nanoemulsions as antifungal, antibacterial and plant growth stimulators.

2. Thymol Nanoemulsion: Methods of Preparation

Techniques for the preparation of nanoemulsions are broadly divided into two categories: High energy approaches and low energy approaches (Fig. 2). The high energy approaches disrupt the oil and aqueous phases into tiny droplets using mechanical devices such as high-pressure homogenizers, microfluidizers and ultrasonicators (Mason 2012). In the case of low energy methods, emulsion is formed spontaneously by mixing all the ingredients (oil, water, surfactant and/or co-surfactant) together. Low energy methods include spontaneous emulsification (SE) method, emulsion inversion point (EIP) method, phase inversion composition (PIC) method and phase inversion temperature (PIT) method. Using these methods, droplet size within mixed oil-water emulsifier systems can be reduced by varying emulsion composition and altering the solution or environmental factors like temperature (Ghosh et al. 2013). High energy processes are too energy intensive to be applied on an industrial scale but can reduce droplet size to submicron range. Low energy emulsification methods require a large amount of surfactant, very precise control of physico-chemical parameters and are not applicable to large-scale industrial processes. Besides, they are more prone to coalescence and creaming (Tadros et al. 2004).

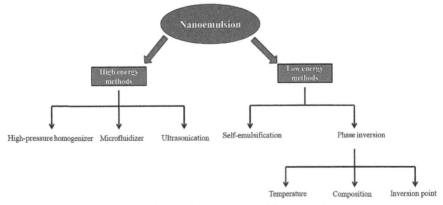

Fig. 2. Various methods of nanoemulsion preparation.

2.1 High-Energy Approaches

High-energy approaches are one of the handiest means of producing nanoemulsions because they can be used in a range of different oils and emulsifier types and are capable of large-scale production. The particle size produced by high energy approaches is managed by a balance between two divergent processes occurring within homogenizer-droplet disruption and droplet coalescence. Mechanical devices, such as high-pressure homogenizer, microfludizer and ultrasonic devices, which are proficient in generating tremendously intense disruption forces, are capable of producing the tiny droplets necessary to form nanoemulsion.

2.1.1 High-Pressure Homogenizer

High-pressure valve homogenizer is probably the most common method for the production of conventional emulsions with small droplet sizes (Schubert and Engel 2004). It is more effective in reducing the size of droplets in pre-existing coarse emulsions than in creating emulsions directly from two separate liquids. In nanoemulsion production, the coarse emulsions produced using rotor-stator systems are fed directly into the inlet of a high-pressure valve homogenizer. Then, the emulsion is pulled into a chamber by the backstroke of a pump and forced through a narrow valve present at the end of the chamber on its forward stroke (Tadros et al. 2004). As the coarse emulsion passes through the narrow valve, the large droplets break down into smaller ones by a combination of the intense disruptive forces, such as shear stress, cavitation and turbulent flow conditions, acting on them (Stang et al. 2001) (Fig. 3). In high-pressure valve homogenizer, the droplet size usually decreases as the number of passes and/or the homogenization pressure increases. A number of recent studies have used high-pressure homogenization to produce

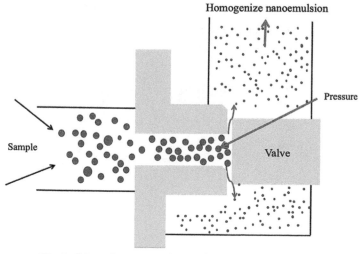

Fig. 3. Schematic representation of a high-pressure homogenizer.

thymol-based nanoemulsions which could potentially be used in the food industry (Ziani et al. 2011, Chang et al. 2012, Xue et al. 2013, Pan et al. 2014, Chen et al. 2015, Chang et al. 2015, Su et al. 2016, Ma et al. 2016, Xue et al. 2017).

2.1.2 Microfluidizer

Microfluidizer is fairly similar in design to a high-pressure homogenizer, but the design of the channel for the flow of emulsion is different. It involves the application of high pressure on a coarse emulsion for the production of nanoemulsions. Microfluidizer produces smaller droplets and narrower distribution of emulsion droplet size (EDS) as compared to traditional emulsification techniques (Pinnamaneni et al. 2003). "Interaction chamber" in microfluidizer is the heart of this device and works on the principle of separating an emulsion flowing through a channel into two streams, passing each stream through a separate fine channel and then directing the two streams at each other in the interaction chamber. Intense disruptive forces are produced within the interaction chamber when the two fast-moving streams of emulsion interrupt upon each other and form extremely efficient droplet disruption (McClements and Rao 2011) (Fig. 4). The process stream is distributed by a pneumatically powered pump that is capable of pressurizing the in-house compressed air (150–650 kPa) up to about 150 MPa. This method has not been used for thymol nanoemulsion preparation till date.

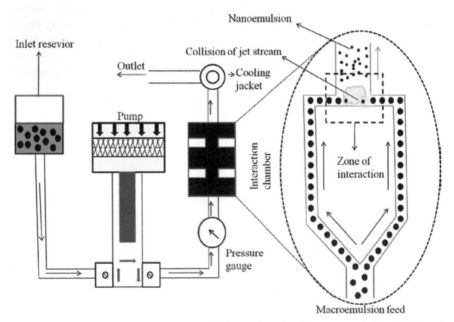

Fig. 4. Schematic representation of a microfludizer (Adopted and reconstructed from Singh et al. 2017).

2.1.3 Ultrasonicator

Ultrasound method is an efficient technique for the preparation of nanoemulsions and has better control over the particle size distribution and stability of nanoemulsions. In this method, ultrasound waves are transferred through the liquid medium where they create a cavitation phenomenon. This causes the formation, growth and implosive collapse of microbubbles/cavities which aid in breaking up primary droplets of dispersed oil into droplets of submicron size. Emulsification by ultrasound method usually occurs in two ways, first droplet generation in the acoustic field and second creation of intense turbulence during asymmetric cavity collapse which causes the break-up and dispersion of droplets in the continuous phase (Li and Fogler 1978b) (Figs. 5 and 6). Emulsification can be achieved using ultrasound in the range of 16–100 kHz (Canselier et al. 2002) either as a batch or continuous process. Size of the droplets produced using an ultrasonic device tends to decrease as the intensity of the ultrasonic waves is increased or the residence time in the disruption zone is increased. The homogenization efficiency also depends on the type and amount of emulsifier present and viscosity of the oil and aqueous phases (Leong et al. 2009). Moghimi et al. (2016) produced nanoemulsion (diameter = 143 nm) of thyme oil extracted from *T. daenensis* by using Tween 80 as surfactant using high intensity ultrasound method. Thymol nanoemulsion has also been prepared by ultrasonication using saponin as the surfactant (Fig. 7). Developed method opens up a new avenue for making nanoemulsion using plant-based components for its use in agriculture system (Kumari et al. 2018).

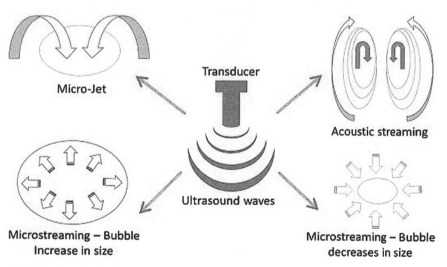

Fig. 5. Scheme of cavitation phenomenon in ultrasonication. (The figure is adopted from Kadam et al. 2015 with copyright permission from Elsevier.)

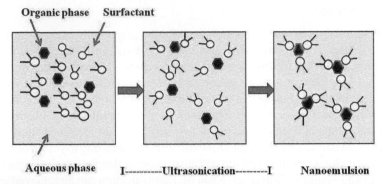

Organic phase   Surfactant

Aqueous phase   I----------Ultrasonication--------I   Nanoemulsion

Fig. 6. Process of emulsification by ultrasonication method for nanoemulsion preparation.

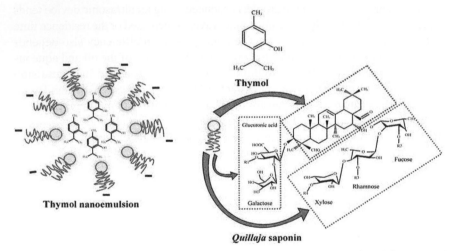

Thymol

Thymol nanoemulsion

Quillaja **saponin**

Fig. 7. Structure of nanoemulsion droplets of thymol nanoemulsion. (Figure adopted from Kumari et al. 2018 with copyright permission from Nature publisher.)

2.2 Low-Energy Approaches

In low-energy methods, formation of emulsions relies on the spontaneous formation of oil droplets in oil-water emulsifier mixtures using the internal chemical energy of the system (Tadros et al. 2004). Low-energy methods are cost-effective as compared to high-energy methods, but it is not possible here to use protein or polysaccharide as a emulsifier, instead it requires high concentrations of synthetic surfactants to form nanoemulsions which limit their use in food and agriculture application (McClements and Rao 2011). Ostwald ripening is also seen in nanoemulsions formed using these methods. Low-energy approaches commonly include spontaneous emulsification and phase inversion methods, and have so far not been exploited for the preparation of thymol nanoemulsion.

2.2.1 Self-Emulsification Method

In self-emulsification method, when two liquids (surfactant and/or solvent molecules) are mixed together through rapid diffusion from the dispersed phase to continuous phase with the help of chemical energy released by dilution process, an emulsion/nanoemulsion is formed spontaneously (Ghai and Sinha 2012). Nanoemulsion in this method can be formed in many ways, such as by varying the composition of two phases, by varying the environmental conditions (temperature, pH, and ionic strength), and by varying the mixing conditions (stirring speed, rate of addition and order of addition) (Anton et al. 2009). In this method, dilution can be carried out by adding water to an organic phase containing hydrophobic oil, water-miscible organic solvent, and surfactant, or by slowly adding hydrophobic oil, a hydrophilic surfactant, and a water-miscible organic solvent to water (Anton et al. 2009) (Fig. 8). There is no report on spontaneous emulsification approach where protein or polysaccharide has been used as an emulsifier.

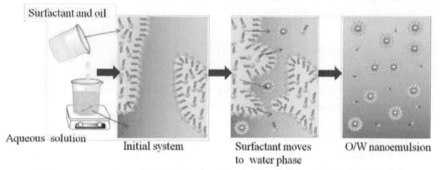

Fig. 8. Schematic representation of proposed mechanism of spontaneous emulsification method. (The figure is adopted from David Julian McClements 2011 with copyright permission from Royal Society of Chemistry.)

2.2.2 Phase Inversion Methods

Nanoemulsion in phase inversion methods are formed by inducing phase inversion in emulsion from a W/O to O/W form or vice versa. Various inversion methods can be induced by changing the temperature (phase inversion temperature, PIT), composition (phase inversion composition, PIC), and emulsion inversion point (EIP). In this method, during the emulsification process, chemical energy is released which is utilized to form nanoemulsion (McClements and Rao 2011).

2.2.2.1 Phase Inversion Temperature (PIT)

This method relies on the changes caused in the physico-chemical properties of non-ionic surfactants with a change in temperature (Anton and Vandamme 2009). Nanoemulsion in PIT method can be formed spontaneously by varying the temperature-time profile of certain mixtures of water and non-ionic surfactant. At low temperatures, surfactant is more soluble in water because the head group of a non-ionic surfactant is highly hydrated. As the temperature increases, solubility of surfactant in water decreases and the head group becomes gradually dehydrated. However, at a particular temperature, solubility of surfactant in oil and water phases is approximately equal. But at higher temperatures, the surfactant becomes more soluble in oil phase than in water phase (McClements and Rao 2011).

In PIT method, preparation of nanoemulsions requires the sample to be brought to its phase inversion temperature (PIT) or hydrophile-lipophile balance (HLB) temperature. At this temperature, extremely low interfacial tension can be achieved in order to promote emulsification because balance exists between hydrophilic and lipophilic properties of the system and the barriers that oppose coalescence processes are low that make the emulsion unstable. Consequently, nanoemulsions can be obtained. Hence, once the small droplets are formed, the temperature has to be immediately shifted from the HLB temperature either by rapid cooling or by rapid heating. This essential step in the PIT method gives rise to kinetically stable nanoemulsions (Solans and Sole 2012).

2.2.2.2 Phase Inversion Composition (PIC)

This method is fairly similar to the PIT method but the change in surfactant spontaneous curvature can be achieved by varying the composition while keeping the temperature constant. This method involves the progressive addition of one of the components, either water or oil, to a mixture of the other two components, oil/surfactant or water/surfactant, respectively (Machado et al. 2012). The PIC method has potential advantages over the PIT method for handling components with temperature-stability problems or systems containing surfactants that are not of the polyoxyethylene type (Solans and Sole 2012).

2.2.2.3 Emulsion Inversion Point Methods

In PIC or PIT methods of emulsification, change is brought about in the properties of surfactant by changing factors such as temperature, pH, or ionic strength of system that leads to transitional-phase inversion from one type of emulsion to another. However, in emulsion inversion point (EIP) methods, change from one type of emulsion to another (e.g., W/O to O/W or vice versa) is through a catastrophic-phase inversion. Catastrophic-phase inversion occurs when the surfactant properties remain constant while the ratio of oil-to-water phase is altered by increasing (or decreasing) the volume fraction of the dispersed phase in an emulsion above or below some critical level (Thakur et al. 2008).

In EIP method, a W/O emulsion containing water droplets dispersed in oil is formed initially using a small molecule surfactant that can stabilize both W/O emulsions (at least for short term) and O/W emulsions (for long term). To achieve phase inversion from a W/O to O/W system, increasing amounts of water are added to the system with constant stirring until the critical point is exceeded. At this stage, the coalescence rate of water droplets exceeds the coalescence rate of oil droplets and leads to phase inversion. The value of critical concentration where phase inversion occurs and the size of oil droplets produced depends on process variables, such as stirring speed, rate of water addition and emulsifier concentration (Thakur et al. 2008). The major factors contributing to emulsification process are the critical surfactant concentration and the surfactant-to-oil ratio (Fernandez et al. 2004).

3. Nanotechnology Intervention in Thymol for Higher Bioactivity

Natural essential oils can be used in agriculture but they face some problems, such as (i) fast degradation (ii) limited water solubility and (iii) short term availability due to their volatile character (Shoji and Nakashima 2004). Therefore, increasing their bioactivity and stability using different strategies has always been considered crucial. Encapsulation of antimicrobial essential oils and other nutrients within a nano-formulation has obvious advantages, such as controlled and sustained release for long lasting effects (Herreo et al. 2014). Encapsulation technology is a promising tool to make effective use of various ingredients in an unmodified form. Coacervation technology with various proteins, starch, gum arabic, corn zein and polymers, such as cellulose, polyvinyl pyrrolidone, chitosan and angico gum, has been used (Meunier et al. 2006, Liolios et al. 2009, Del Toro-Sanchez et al. 2010, Glenn et al. 2010, Marques 2010, Paula et al. 2010, Ponce Cevallos et al. 2010, Guarda et al. 2011, Xiao et al. 2011). In this context, nanoemulsion technology is being investigated as a potential strategy to enhance solubility, stability, and protection of natural essential oils (Donsi et al. 2011). Thymol could be converted into nano-droplets through entrapment in suitable surfactant (Chang et al. 2012, Chang et al. 2015, Kumari et al. 2018). As a result, it becomes physically and chemically stable in the aqueous medium (Mason et al. 2006, McClements 2011, Chang et al. 2012, Kumari et al. 2018). In thymol nanosphere, thymol molecule would stay in the hydrophobic core of the spheres and only its limited number would be released (probably via diffusion mechanism) into the gel and vaporized. From a biological standpoint, nano-droplets of nanoemulsion could be efficiently absorbed on biological surfaces for efficient and wider biological activities. Nanoemulsions are commonly stabilized by amphiphilic surfactants that are absorbed between water and oil phases. They perform two functions in nanoemulsion: (1) Reduce interfacial tension between oil and water phase (2) decrease the rate of coalescence of the oil droplets by forming a physical, steric and/or electric barrier around them. A range of synthetic surfactants have been used by various researchers for the preparation of nanoemulsions (Chang et al. 2012, Ziani et al. 2011, Ghosh et al. 2013, Bhargava et al. 2015, Salvia-Trujillo et al. 2015, Moghimi et al. 2016). Until now, the available literature about thymol nanoemulsion is summarized in Table 1.

Table 1. Synthesis and characterization of thymol nanoemulsion.

Nanoemulsion Components	Method	Z-average (nm)	PDI	Zeta Potential (mV)	Storage Stability	References
Thymol + Saponin	Ultasonication	274	0.13	−29	3 mon	Kumari et al. 2018
Thymol + Gelatin	Homogenization	257.97	0.32	−21.73	N/A	Xue et al. 2017
Thymol + Lauric arginate + Lecithin	Homogenization	100	N/A	N/A	N/A	Ma et al. 2016
Thymol + Tween-80 + Lecithin	Ultrasonication	143.2	N/A	N/A	6 mon	Moghimi et al. 2016
Thymol + Casein hydrolysates + Sucrose stearate	Homogenization	117	0.29	N/A	60 d	Su et al. 2016
Thyme oil + Tween-80 + Lauric arginate	Homogenization	150	N/A	18	30 d	Change et al. 2015
Thymol/eugenol + Zein casein	Homogenization	200	0.21	N/A	N/A	Chen et al. 2015
Thymol + Sodium caseinate	Homogenization	130	N/A	−22	30 d	Pan et al. 2014
Thymol + Propylene glycol	Homogenization	29	N/A	−30	N/A	Xue et al. 2013
Thymol + Tween-80	Homogenization	1300	N/A	N/A	30 d	Chang et al. 2012
Thymol + Tween-80	Homogenization	N/A	N/A	N/A	N/A	Ziani et al. 2011

N/A: Value not available

4. Applications of Thymol and Thymol Nanoemulsion in Disease Control and Plant Growth

Most of the plant diseases are caused by phytopathogenic microorganisms, like fungi, bacteria and viruses (Kotan et al. 2010). Thus, mitigating crop losses due to diseases is an important consideration in crop production. Rapid and effective control of plant disease in crop cultivation is frequently achieved by use of synthetic insecticides and pesticides. However, these chemicals are associated with undesirable effects due to their slow biodegradation and toxic residues in the environment (Isman 2000). The risk of developing resistance by microorganisms and high cost-benefit ratio are among the other disadvantages of synthetic agrochemicals. Therefore, there has been an increased surge of interest in finding alternative pesticides and antimicrobial compounds, including plant extracts and essential oils/essential oil components of aromatic plants (Kotan et al. 2010). Thymol and thyme oil in different preparations have been used in agriculture to impart protection against fungal, bacterial and pest diseases (Kumari et al. 2018). Besides controlling disease, thymol has also shown potential plant growth promotory activity (Kumari et al. 2018).

4.1 Antibacterial Activity

Thymol showed the highest antimicrobial activity as compared to other essential oils like carvacrol, citronellal, eugenol and terpinen-4-ol (Wattanasatcha et al. 2012). It showed antibacterial activities against *Salmonella typhimurium, Staphylococcus aureus, Vibrio parahaemolyticus, Listeria innocua, Escherichia coli* and *Pseudomonas aeruginosa* (Guarda et al. 2011, Wattanasatcha et al. 2012, Kotan et al. 2013). Thymol is highly active against *Botrytis cinerea*, commonly known as gray mold, the main organism responsible for rot in fresh fruit and vegetables (Martinez-Romero et al. 2007, Navarro et al. 2011). A nanocomposite film of essential oil components like carvacrol (CRV) and thymol (TML) embedded in low-density polyethylene/organically modified montmorillonite (LDPE/OMM) was prepared by Campos-Requena et al. (2015) and evaluated on strawberries inoculated with *Botrytis cinerea*. The film functioned as controlled-release food packaging material and reduced ~ 15% release rate of EOCs as compared to neat LDPE. Additionally, the association of CRV and TML in 50:50 ratio exhibited an *in vivo* synergistic antimicrobial effect against strawberry gray mold, diminishing the IC_{50} for *B. cinerea* to one-third of the initial CRV IC_{50} without altering the organoleptic characteristics of strawberries (Campos-Requena et al. 2015). An antibacterial and plant growth promoting nanoemulsion was formulated using thymol, *Quillaja* saponin, a glycoside surfactant of *Quillaja* tree. The nanoemulsion at the concentrations of 0.01–0.06%, v/v showed substantial *in vitro* growth inhibition of *Xanthomonas axonopodis* pv. *glycine* of soybean (6.7–0.0 log CFU/ml). In pot experiments, seed treatment and foliar application of the nanoemulsion (0.03–0.06%, v/v) significantly lowered the disease severity (DS)

(33.3–3.3%) and increased per cent efficacy of disease control (PEDC) (54.9–95.4%) of bacterial pustule in soybean caused by *X. axonopodis* pv. *glycine*. Subsequently, a significant enhancement of plant growth characters was also recorded in plants treated with thymol nanoemulsion (Kumari et al. 2018).

4.2 Antifungal Activity

Thymol has been reported as one of the most potent antifungal compounds against fungi such as *Penicillium digitatum*, *Penicillium italicum*, *Aspergillus niger*, *Rhizoctonia solani*, *Fusarium oxysporium*, *Penecillium digitatum*, etc. (Ahmad et al. 2011, Marei et al. 2012, Janatova et al. 2015). The antifungal efficacy of thymol and carvacrol was reported on the growth of *P. digitatum* and *P. italicum* in liquid media and PDA plates. These molecules also reduced the infected surface on lemon caused by *P. digitatum*, indicating their usefulness as an alternative for controlling postharvest lemon decay by incorporating in wax usually applied in the packing lines. The best results were obtained with thymol and, at the concentration of 250 μl l^{-1}, the development of *P. digitatum* on lemon surface was totally inhibited (Perez-Alfonso et al. 2012). The potential of thyme essential oil in controlling gray mold and *Fusarium* wilt and inducing systemic acquired resistance in tomato seedlings and tomato grown in hydroponic system was evaluated by Ben-Jabeur et al. (2015). Thyme oil reduced 64% of *Botrytis cinerea* colonization on pretreated detached leaves as compared to untreated control. It also decreased the *Fusarium* wilt severity to the extent of 30.76% at 7 days post treatment (Ben-Jabeur et al. 2015). A sub-micron thymol-derived emulsion was prepared by Gill et al. (2016) which was effective against *Fusarium graminearum*, a very destructive fungal pathogen causing Fusarium head blight (FHB) in wheat. Spraying of thymol emulsion on wheat heads after inoculation reduced the infection rate. Both thymol and thymol emulsion were effective in controlling FHB and did not show any cytotoxic effect at applied concentration of 0.1% indicating that thymol-based emulsion can be a promosing candidate for a naturally derived fungicide with a low environmental footprint.

4.3 Plant Growth and Defense System Promoting Activity

Thymol extracted from *Satureja hortensis* was used as seed disinfectant in lettuce and tomato seeds against plant pathogenic bacteria. It decreased the disease severity rate to minimum level and increased germination per cent of seeds, plant heights, shoots and roots fresh weights in pot assays (Kotan et al. 2013). An antibacterial and plant growth promoting nanoemulsion was formulated using thymol and *Quillaja* saponin and evaluated for its effect on soybean. Among the various concentrations of nanoemulsion, 0.03% concentration was found most effective in increasing germination percentage, shoot-root length, root and fresh weight as compared to controls in soybean seedling growth experiments (Kumari et al. 2018). Thymol induced changes in root morphology to make the plant more resistant towards pathogens (Tahat et al. 2011) and significantly increased the stem length, dry weight and biomass of tomato as compared to infected control, causing a great increase in yield (Abo-Elyousr et al. 2014). Application of thyme oil induced accumulation of phenolic compounds in roots and peroxidase activity in leaves and roots. It is also involved in cell wall strengthening and lignifications in plant cell that limit the microbial penetration. Apparently, peroxidase accumulation is the first defense activation signal which might be attributed to the biological effect of essential oils and their components causing oxidative burst by accumulation of reactive oxygen species (ROS). Actually, ROS have a range of effects on plant defense responses, including defence gene activation, as well as defence compounds stimulation (Levine et al. 1994). All these findings suggest that thymol-based nanoemulsion can be a potential antimicrobial, plant growth and defense system promoting agent in agriculture.

5. Modes of Antimicrobial Action

The antimicrobial activity spectrum observed for many EO nanoemulsions is often broader and stronger than that of free EOs. They possess bactericidal action against a variety of pathogenic microorganisms and their action involves various mechanisms in the cell. The possible mechanisms for essential oils and their components against bacterial cells as reported in the literature are (1) degradation of cell wall (2) damage to cytoplasmic membrane (3) damage to membrane proteins (4) leakage of cell contents and (5) coagulation of cytoplasm and depletion of proton motive force (Burt 2004). Thymol nanoemulsion droplets having negative surface charge have been developed and their adhesion with bacterial cell wall components probably occur through self-assembly or chemical linkage rather than electrostatic interaction which causes disintegration of the cell wall and cytoplasmic membranes, and ultimately release cellular constituents (Majeed et al. 2016, Kumari et al. 2018). As lipophilic agents, essential oils execute their action at the level of membrane and membrane embedded enzymes. Their action is due to a change in the fatty acid composition of the cell membrane. Thymol has been reported to show fungicidal activity by inhibition of ergosterol biosynthesis and disruption of membrane integrity (Ahmad et al. 2011). These effects may eventually result in changes such as inhibition of respiration and alteration in permeability. Thymol is a potent inhibitor of pectin methylesterase (PME), an enzyme that modifies the degree of methylesterification of pectin, a major component of fungal cell walls. Such changes in pectin structure are associated with changes in cellular adhesion, plasticity, pH and ionic contents of the cell wall and influence fungi development, membrane integrity and permeability. On the other hand, fungi produce cellulase to degrade cell walls during pathogenesis and inhibition of this enzyme by thymol ultimately affects the disease development. Thymol exhibits strong inhibitory effect on cellulase activity (Marei et al. 2012). In general, the nanoemulsion-based delivery of EOs interacts with the microbial cell membranes through mainly four routes, as represented in Fig. 9.

1. The nanoemulsion droplets, having increased surface area with hydrophilic surface and passive transport through the outer cell membrane, improve the interaction with cell membrane via the abundant porin proteins and disturb the phospholipid bilayer membrane integrity.
2. Nanoemulsion droplets with emulsifier fuse with the phospholipid bilayer membrane of the cell membrane and promote the release of EOs at the target sites.
3. The sustained release of EOs from nanoemulsion droplets over time, driven by EO partition between the oil droplets and aqueous phase, prolong the activity of EOs.
4. The electrostatic interactions of positively charged nanoemulsion droplets with negatively charged microbial cell membranes increase the concentration of EOs at the site of action. In case of nanoemulsion droplet with negative surface charge, interactions with microbial cell membranes probably occur through self assembly or chemical linkage rather than electrostatic interactions.

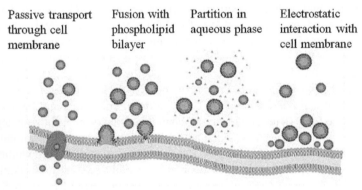

Fig. 9. Schematic representation of different routes adopted by nanoemulsion for interactions of EOs with microbial cell membranes. The figure is adopted from Donsi et al. (2016) with copyright permission from Elsevier.

Current and Future Prospects of Thymol Nanoemulsions

Strong antimicrobial activity of thymol makes this molecule a very voluble compound to be used in various products and processes related to human use. Thus, thymol has been exploited in daily use products, like mouth wash, tooth paste and other pharmaceutical products. With the advancement of process and extraction technology, the thymol becomes comparatively economical to other plant essential oil components (EOCs). Hence, we foresee that thymol could be exploited in agriculture for crop disease management. With the use of nanotechnology, thymol can be easily exploited in nanoscale for many more applications in the agricultural field. Translation of thymol into thymol nanoformulation can enhance its use efficacy and use in lower concentrations. Development of thymol nanoemulsion-based delivery system is an innovative, novel, forthcoming and eco-friendly approach for various compounds of agricultural importance, like micronutrients, vitamins, food preservatives, etc. Thymol has GRAS food ingredient status, thus, it could be applied and used in food crops, particularly in postharvest for controlling fruit decay and as natural food preservative. Overall, thymol-based nanoemulsions open up a new avenue for development of bio-based carrier for delivery of bioactive compounds/agrochemicals for their efficient use in agriculture. Nanoemulsion could be an alternative to synthetic agrochemicals as an ecofriendly approach for sustainable agriculture and protection of the biosphere.

References

Abo-Elyousr, K.A.M., M.A.A. Seleim, K.M.H. Abd-El-Moneem and F.A. Saead. 2014. Integrated effect of *Glomus mosseae* and selected plant oils on control of tomato bacterial wilt. Crop Prot. 66: 67–71.

Ahmad, A., A. Khan. F. Akhtar. S. Yousuf, I. Xess and L.A. Khan. 2011. Fungicidal activity of thymol and carvacrol by disrupting ergosterol biosynthesis and membrane integrity against Candida. Eur. J. Clin. Microbiol. Infect. Dis. 30: 41–50.

Anton, N. and T.F. Vandamme. 2009. The universality of low-energy nanoemulsification. Int. J. Pharmaceut. 377(1): 142–147.

Bhargava, K., D.S. Conti, S.R. da Rocha and Y. Zhang. 2015. Application of an oregano oil nanoemulsion to the control of foodborne bacteria on fresh lettuce. Food Microbiol. 47: 69–73.

Burt, S. 2004. Essential oils: Their antibacterial properties and potential applications in foods—a review. Int. J. Food Microbiol. 94: 223–253.

Campos-Requena, V.H., B.L. Rivas, A. Perez, C.A. Figueroa and E.A. Sanfuentes. 2015. The synergistic antimicrobial effect of carvacrol and thymol in clay/polymer nanocomposite films over strawberry gray mold. LWT—Food Sci. Technol. 64: 390–396.

Canselier, J.R., H. Delmas, A.M. Wilhelm and B. Abismail. 2002. Ultrasound emulsification—An overview. J. Disper. Sci. Technol. 23: 333–349.

Chang, Y., L. McLandsborough and D.J. McClements. 2012. Physical properties and antimicrobial efficacy of thyme oil nanoemulsions: Influence of ripening inhibitors. J. Agric. Food Chem. 60(48): 12056–12063.

Chang, Y., L. McLandsborough and D.J. McClements. 2015. Fabrication, stability and efficacy of dual-component antimicrobial nanoemulsions: Essential oil (thyme oil) and cationic surfactant (lauric arginate). Food Chem. 172: 298–304.

Chen, H., Y. Zhang and Q. Zhong. 2015. Physical and antimicrobial properties of spray-dried zein–casein nanocapsules with coencapsulated eugenol and thymol. J. Food Eng. 144: 93–102.

da Silva, M.A., E. da Daemona, C.M.O. Monteiro, R. Maturanoa, F.C. Britoa and T. Massoni. 2011. Acaricidal activity of thymol on larvae and nymphs of *Amblyomma cajennense* (acari: Ixodidae). Vet. Parasitol. 183: 136–139.

Del Toro-Sanchez, C., J. Ayala-Zavala, L. Machi, H. Santacruz, M. Villegas-Ochoa, E. Alvarez-Parrilla and G. Gonzalez-Aguilar. 2010. Controlled release of antifungal volatiles of thyme essential oil from-cyclodextrin capsules. J. Incl. Phenom. Macrocycl. Chem. 67: 431–441.

Di Pasqua, R.D., G. Betts, N. Hoskins, M. Edwards, D. Ercolini and G. Mauriello. 2007. Membrane toxicity of antimicrobial compounds from essential oils. J. Agric. Food Chem. 55: 4863–4870.

Donsi, F., M. Annunziata, M. Sessa and G. Ferrari. 2011. Nanoencapsulation of essential oils to enhance their antimicrobial activity in foods. LWT—Food Sci. Technol. 44(9): 1908–1914.

Dinsi, F. and G. Ferrari. 2016. Essential oil nanoemulsions as antimicrobial agents in food. J. Biotechnol. 233: 106–120.

Esmaeili, A. and A. Asgari. 2015. *In vitro* release and biological activities of *Carumcopticum* essential oil (CEO) loaded chitosan nanoparticles. Int. J. Biol. Macromol. 81: 283–290.

Evans, J.D. and S.A. Martin. 2000. Effects of thymol on ruminal microorganisms. Curr. Microbiol. 41: 336–340.

Fernandez, P.V., Andre. J. Rieger and A. Kuhnle. 2004. Nanoemulsion formation by emulsion phase inversion. Colloid Surf. A 251(1): 53–58.

Ghai, D. and V.R. Sinha. 2012. Nanoemulsions as self-emulsified drug delivery carriers for enhanced permeability of the poorly water-soluble selective β 1-adrenoreceptor blocker Talinolol. Nanomed-Nanotechnol. 8(5): 618–26.

Ghosh, V., S. Saranya, A. Mukherjee and N. Chandrasekaran. 2013. Cinnamon oil nanoemulsion formulation by ultrasonic emulsification: Investigation of its bactericidal activity. J. Nanosci. Nanotechnol. 13: 114–122.

Gill, T., J. Li, M. Saenger and S. Scofield. 2016. Thymol-based submicron emulsions exhibit antifungal activity against *Fusarium graminearum* and inhibit fusarium head blight in wheat. J. Appl. Microbiol. 121: 1103–1116.

Glenn, G.M., A.P. Klamczynski, S.H. Imam, B. Chiou, W.J. Orts and D.F. Woods. 2010. Encapsulation of plant oils in porous starch microspheres. J. Agric. Food Chem. 58: 4180–4184.

Guarda, A., J.F. Rubilar, J. Miltz and M.J. Galotto. 2011. The antimicrobial activity of microencapsulated thymol and carvacrol. Int. J. Food Microbiol. 146: 144–150.

Herrero, A.M., P. Carmona, F. Jimenez-Colmenero and C. Ruiz-Capillas. 2014. Polysaccharide gels as oil bulking agents: Technological and structural properties. Food Hydrocoll. 36: 374–381.

Isman, M.B. 2000. Plant essential oils for pest and disease management. Crop Prot. 19: 603–608.

Janatova, A., A. Bernardos, J. Smid, A. Frankova, M. Lhotka, L. Kourimska, J. Pulkrabek and P. Kloucek. 2015. Long-term antifungal activity of volatile essential oil components released from mesoporous silica materials. Ind. Crops Prod. 67: 216–220.

Kotan, R., A. Cakir, F. Dadasoglu, T. Aydin, R. Cakmakci, H. Ozer, S. Kordali, E. Mete and N. Dikbas. 2010. Antibacterial activities of essential oils and extracts of Turkish Achillea, Satureja and Thymus species against plant pathogenic bacteria. J. Sci. Food Agric. 90: 145–160.

Kotan, R., F. Dadasoglu, K. Karagozb, A. Cakirc, H. Ozerd, S. Kordalia, R. Cakmakcid and N. Dikbase. 2013. Antibacterial activity of the essential oils and extracts of *Satureja hortensis* against plant pathogenic bacteria and their potential use as seed disinfectants. Sci. Hortic. 153: 34–41.

Kumari, S., R.V. Kumaraswamy, R.C. Choudhary, S.S. Sharma, A. Pal, R. Raliya, P. Pratim Biswas, and V. Saharan. 2018. Thymol nanoemulsion exhibits potential antibacterial activity against bacterial pustule disease and growth promotory effect on soybean. Sci. Rep. 8: 6650.

Lambert, R., P.N. Skandamis, P.J. Coote and G.J. Nychas. 2001. A study of the minimum inhibitory concentration and mode of action of oregano essential oil, thymol and carvacrol. J. Appl. Microbiol. 91: 453–462.

Levine, A., R. Tenhaken, R. Dixon and C. Lamb. 1994. H_2O_2 from the oxidative burst orchestrates the plant hypersensitive disease resistance response. Cell. 79: 583–593.

Leong, T.S.H., T.J. Wooster, S.E. Kentish and M. Ashokkumar. 2009. Minimising oil droplet size using ultrasonic emulsification. Ultrason. Sonochem. 16(6): 721–727.

Li, M.K. and H.S. Fogler. 1978b. Acoustic emulsification. Part 2. Break-up of the larger primary oil droplets in a water medium. J. Fluid Mech. 88(3): 513–528.

Liolios, C.C., O. Gortzi, S. Lalas, J. Tsaknis and I. Chinou. 2009. Liposomal incorporation of carvacrol and thymol isolated from the essential oil of *Origanum dictamnus* L. and *in vitro* antimicrobial activity. Food Chem. 112: 77–83.

Ma, Q., P.M. Davidson and Q. Zhong. 2016. Antimicrobial properties of microemulsions formulated with essential oils, soybean oil, and Tween 80. Int. J. Food Microbiol. 226: 20–25.

Machado, A.H., D. Lundberg, A.J. Ribeiro, F.J. Veiga, B. Lindman, M.G. Miguel and U. Olsson. 2012. Preparation of calcium alginate nanoparticles using water-in-oil (w/o) nanoemulsions. Langmuir. 28(9): 4131–4141.

Majeed, H., L. Fei, H. Joseph, R.S. Hafiz, Q. Jing, A. Barkat, B. Yuan-Yuan, M. Jianguo, Y. Wallace and Z. Fang. 2016. Bactericidal action mechanism of negatively charged food grade clove oil nanoemulsions. Food Chem. 197: 75–83.

Marei, G.I.K., M.A. Abdel Rasoul and S.A.M. Abdelgaleil. 2012. Comparative antifungal activities and biochemical effects of monoterpenes on plant pathogenic fungi. Pesticide Biochemistry and Physiology 103: 56–61.

Marques, H.M.C. 2010. A review on cyclodextrin encapsulation of essential oils and volatiles. Flavour Fragr. J. 25: 313–326.

Martinez-Romero, D., F. Guillen, J.M. Valverde, G. Bailen, P. Zapata, M. Serrano, S. Castillo and D. Valero. 2007. Influence of carvacrol on survival of *Botrytis cinerea* inoculated in table grapes. Int. J. Food Microbiol. 115: 144–148.

Mason, T.G., J.N. Wilking, K. Meleson, C.B. Chang and S.M. Graves. 2006. Nanoemulsions: Formation, structure, and physical properties. J. Phys. Condens. Matter. 18(41): 635–666.

McClements, D.J. and J. Rao. 2011. Food-grade nanoemulsions: Formulation, fabrication, properties, performance, biological fate and potential toxicity. Crit. Rev. Food Sci. 51(4): 285–330.

Meunier, J.P., J.M. Cardot, P. Gauthier, E. Beyssac and M. Alric. 2006. Use of rotary fluidized-bed technology for development of sustained-release plant extracts pellets: Potential application for feed additive delivery. J. Anim. Sci. 84: 1850–1859.

Moghimi, R., L. Ghaderi, H. Rafati, A. Aliahmadi and D.J. McClements. 2016. Superior antibacterial activity of nanoemulsion of *Thymus daenensis* essential oil against *E. coli*. Food Chem. 194: 410–415.

Navarro, D., H.M. Diaz-Mula, F. Guillen, P.J. Zapata, S. Castillo and M. Serrano. 2011. Reduction of nectarine decay caused by *Rhizopus stolonifer, Botrytis cinerea and Penicillium digitatum* with Aloe vera gel alone or with the addition of thymol. Int. J. Food Microbiol. 151: 241–246.

Oussalah, M., S. Caillet, L. Saucier and M. Lacroix. 2006. Antimicrobial effects of selected plant essential oils on the growth of a *Pseudomonas putida* strain isolated from meat. Meat Sci. 73: 236–244.

Pan, K., H. Chen, P.M. Davidson and Q. Zhong. 2014. Thymol nanoencapsulated by sodium caseinate: Physical and antilisterial properties. J. Agric. Food Chem. 62(7): 1649–1657.

Paula, H.C.B., F.M. Sombra, F.O.M.S. Abreu and R.C.M. Paul. 2010. Lippia sidoides essential oil encapsulation by angico gum/chitosan nanoparticles. Braz. J. Chem. Eng. 21: 2359–2366.

Pinnamaneni, S., N.G. Das and S.K. Das. 2003. Comparison of oil-in water emulsions manufactured by Microfluidization and homogenization. Pharmazie. 58(8): 554–558.

Ponce Cevallos, P.A., M.P. Buera and B.E. Elizalde. 2010. Encapsulation of cinnamon and thyme essential oils components (cinnamaldehyde and thymol) in cyclodextrin: Effect of interactions with water on complex stability. J. Food Eng. 99: 70–75.

Poulose, A.J. and R. Croteau. 1978. Biosynthesis of aromatic monoterpenes: Conversion of γ-terpinene to *p*-cymene and thymol in *Thymus vulgaris* L. Arch. Biochem. Biophys. 187: 307–314.

Perez-Alfonso, C.O., D. Martínez-Romero, P.J. Zapata, M. Serrano, D. Valero and S. Castillo. 2012. The effects of essential oils carvacrol and thymol on growth of *Penicillium digitatum* and *P. italicum* involved in lemon decay. Int J. Food Microbiol. 158: 101–106.

Sacchetti, G., S. Maietti, M. Muzzoli, M. Scaglianti, S. Manfredini, M. Radice and R. Bruni. 2005. Comparative evaluation of 11 essential oils of different origin as functional antioxidants, antiradicals and antimicrobials in foods. Food Chem. 91: 621–632.

Salvia-Trujillo, L., A. Rojas-Grau, R. Soliva-Fortuny and O. Martin-Belloso. 2015. Physicochemical characterization and antimicrobial activity of food-grade emulsions and nanoemulsions incorporating essential oils. Food Hydrocoll. 43: 547–556.

Schubert, H. and R. Engel. 2004. Product and formulation engineering of emulsions. Chem. Eng. Res. Des. 82(9): 1137–1143.

Shapira, R. and E. Mimran. 2007. Isolation and characterization of *Escherichia coli* mutants exhibiting altered response to thymol. Microb. Drug Resist. 13: 157–165.

Shoji, Y. and H. Nakashima. 2004. Nutraceutics and delivery systems. J. Drug Target. 12: 385–391.

Singh, Y., J.G. Meher. K. Raval, F.A. Khan, M. Chaurasia, N.K. Jain and M.K. Chourasia. 2017. Nanoemulsion: Concept, development and applications in drug delivery. J. Control. Release. 252: 28–49.

Solans, C. and I. Sole. 2012. Nano-emulsions: Formation by low-energy methods. Curr. Opin. Colloid. In 17(5): 246–254.

Stang, M., H. Schuchmann and H. Schubert. 2001. Emulsification in high-pressure homogenizers. Eng. Life Sci. 1(4): 151–157.

Su, D. and Q. Zhong. 2016. Formation of thymol nanoemulsions with combinations of casein hydrolysates and sucrose stearate. J. Food Eng. 179: 1–10.

Tadros, T., R. Izquierdo, J. Esquena and C. Solans. 2004. Formation and stability of nano-emulsions. Adv. Colloid. Interface Sci. 108-109: 303–318.

Thakur, R.K., C. Villette, J.M. Aubry and G. Delaplace. 2008. Dynamic emulsification and catastrophic phase inversion of lecithin-based emulsions. Colloid Surf. A 315(1): 285–293.

Wattanasatcha, A., S. Rengpipat and S. Wanichwecharungruang. 2012. Thymol nanospheres as an effective anti-bacterial agent. Int. J. Pharm. 434: 360–365.

Xiao, D., P.M. Davidson and Q. Zhong. 2011. Spray-dried zein capsules with coencapsulated nisin and thymol as antimicrobial delivery system for enhanced antilisterial properties. J. Agric. Food Chem. 59: 7393–7404.

Xue, J., P.M. Davidson and Q. Zhong. 2013. Thymol nanoemulsified by whey protein-maltodextrin conjugates: the enhanced emulsifying capacity and antilisterial properties in milk by propylene glycol. J. Agric. Food Chem. 61(51): 12720–12726.

Xue, J., P.M. Davidson and Q. Zhong. 2017. Inhibition of *Escherichia coli* O157:H7 and *Listeria monocytognes* growth in milk and cantaloupe juice by thymol nanoemulsions prepared with gelatin and lecithin. Food Control. 73: 1499–1506.

Ziani, K., Y. Chang, L. McLandsborough and D.J. McClements. 2011. Influence of surfactant charge on antimicrobial efficacy of surfactant-stabilized thyme oil nanoemulsions. J. Agric. Food Chem. 59(11): 6247–6255.

CHAPTER 11

Phyto-Nano Interaction

An Insight to the Phenomenon

Divya Vishambhar Kumbhakar,[1] Debadrito Das,[1]
Bapi Ghosh,[1] Ankita Pramanik,[1] Sudha Gupta[2] and
Animesh Kumar Datta[1,]*

1. Introduction

Nanotechnology is an emerging branch of science enduring wide applications in industry (Roco 2003), medicine (Masarovičová and Kráľová 2013, Biffi et al. 2015), electronics (Frank et al. 1998, Kong et al. 1999), biotechnology (Scrinis and Lyons 2007), agriculture (Raliya et al. 2013, Halder et al. 2015a,b) including host pathogen interaction (Nair et al. 2010) and different areas of life sciences (Selivanov and Zorin 2001, Nguyen et al. 2015, Benckiser et al. 2017) among others. However, due to frequent release of nanoparticles (NPs—at least one dimension ranging between 1–100 nm—(Roco 2003) possess unique physico-chemical properties (Remédios et al. 2012, Masarovičová and Kráľová 2013)) in the environmental inter-collegium by natural processes as well as by anthropogenic activities (Nowack and Bucheli 2007, Tervonen et al. 2009, Lidén 2011) different components of the ecosystem are affected. Furthermore, indiscriminate use of such nanomaterials results in release of nanowaste which causes toxicity (Bahadar et al. 2016, Liu et al. 2016, Di Bucchianico et al. 2017, Dumala et al. 2017); this is of serious global concern.

Plant species are the primary component of the ecosystem, specifically, rooted plants (as soil serves as a sink for deposited NPs in the environment, facilitating nano-bio interaction), due to their operational simplicity and cost effectivity, can

[1] Department of Botany, Cytogenetics, Genetics and Plant Breeding Section, University of Kalyani, Kalyani-741235, West Bengal, India.
[2] Department of Botany, Pteridology and Palaeobotany Section, University of Kalyani, Kalyani-741235, West Bengal, India.
* Corresponding author: dattaanimesh@gmail.com

be used as model for monitoring nanotoxicity (Kumbhakar et al. 2016a, Ghosh et al. 2017a). International Programme on Chemical Safety (IPCS, WHO) and United Nation Environment Programme (UNEP) also authenticated the significance of using plant species for assessment of the toxicological impact of nanomaterials on environmental components.

Physical properties and chemoreactivity of the released nanoparticles are determinant factors for their bioactivity (Padmavathy and Vijayaraghavan 2008, Yang et al. 2017). NPs are reported to cause cytotoxic and genotoxic effects on mitotic cells of different plant species (Kumbhakar et al. 2016a, Ghosh et al. 2017a, Pramanik et al. 2018) but changes in mitotic cells (Kumbhakar et al. 2017a) have significant consequences relating to heritable changes in subsequent generations. Furthermore, release of NPs in the environmental inter-collegium interact with the cellular systems of plant species and can induce stress accumulation in a host system, thereby activating their own defence mechanism by triggering anti-oxidant enzymes. Such studies in different plant species highlight differential defence activation by the host (Das et al. 2017a, Ghosh et al. 2018). Persistence of stress environment due to NPs accumulation induces genotoxicity in the form of DNA double strand break as evinced from comet assay. DNA double strand break can be repaired and fixed into mutation. Nano-bio interaction can bring about cellular mortality (studied from fluorescence assisted cell sorting) which is significant for dose monitoring and suitable applicability of engineered nanoparticles. Therefore, assessment of plant-nano interaction is significant for the proper exploration of nanomaterials in a wider perspective. With the view to it, the present article describes the types, synthesis, characterization, exposure, uptake, localization and transport of NPs in a plant system, inducing biological effects like physiological (stress induction and subsequent antioxidant responses and apoptotic cell death), cytogenetical (mitotic and meiotic aberrations) and genotoxic (DNA damage) effects. The application of NPs in crop improvement is also highlighted.

2. Types of Nanoparticles

Nanomaterials can be classified either based on chemical nature of the nano element/compound or by their structural geometry.

2.1 Classification Based on Chemical Nature of NPs

Under this categorization, nanoparticles are subdivided into three major types, as follows:

2.1.1 Non-Metallic NPs

Such NPs types are synthesized from the non-metallic elements and they are carbon nanostructures (Mostofizadeh et al. 2011), fullerenes (Georgakilas et al.

2015), liposomes (Ali et al. 2016) and dendrimers (Boal et al. 2002, Frankamp et al. 2005), among others.

2.1.2 Metallic NPs

Such nanostructures reside in their pure metallic forms and are exemplified as copper (Cu), iron (Fe), stannum (Sn), etc. However, metallic structures are very difficult to maintain in their pure elemental form as they are prone to oxidation (Wu et al. 2008).

2.1.3 Semiconductor NPs

With a switchable band gap, semiconductor NPs are the most frequent and wide class of nanomaterials manufactured so far. The broad class includes zinc oxide (ZnO), stannous oxide (SnO-Anjum et al. 2016), ferric oxide (Fe_2O_3-Hao et al. 2015), copper oxide (CuO) and cadmium sulphide (CdS-Kumbhakar et al. 2016a) NPs, among others.

2.2 Classification Based on Particle Geometry

Under the typification subject, nanostructures are broadly classified into following categories:

2.2.1 Zero Dimensional Nanoparticles

Zero dimensional NPs include quantum dots, heterogeneous particle arrays, core–shell quantum dots, onions, hollow spheres and nanolenses (Zhang and Wang 2008, Lee et al. 2009, Wang et al. 2009, Kim et al. 2010). These are used in LEDs (Stouwdam and Janssen 2008, Singh 2013), solar cells (Lee et al. 2009), single-electron transistors (Mokerov et al. 2001), lasers (Ustinov et al. 2000) and several other applications.

2.2.2 One Dimensional Nanoparticles

One dimensional systems include thin films (1–100 nm) or monolayers (Tiwari et al. 2012) and nanotips (Xu and Wang 2011) and they are the only member under the highly specialized engineered class. This class is made up of the major structural, chemical and biological sensors (Pradhan et al. 2008), magneto-optics (Li et al. 2015), digital storage (Zheludkevicha et al. 2005) and fibre optic systems (Urrutia et al. 2015).

2.2.3 Two Dimensional Nanoparticles

This is the most diversified class, comprising more than 90.0% of the nanoparticles manufactured. Morphological diversities include nanospheres (Liu et al. 2006), nanotubes (Georgakilas et al. 2015), cuboids (Ruso et al. 2013), hexagons (Wang et al. 2010), etc.

2.2.4 Three Dimensional Nanoparticles

This class comprises complex secondary to tertiary geometry, which includes dendrimers (Pan et al. 2010), fullerenes (Bhatia 2016), nanorosette (Das et al. 2017a), tetrapods (Jin et al. 2014, Papavlassopoulos et al. 2014), etc.

3. Synthesis of Nanoparticles

Nanomaterial syntheses are broadly categorized into top-down and bottom-up approaches.

Top-down approaches include mechanical attrition of bulk materials, yielding wide particle size distribution ranging from 10 to 10^3 nm. However, the method is less precise, produces varied particle geometry and contains impurities.

Bottom-up approaches are associated with various physical and chemical techniques, involving production of nanomaterials by controlled aggregation of seed molecules. The domain includes several techniques, namely, wet-chemical co-precipitation (Faraji et al. 2010), thermal evaporation (Kungumadevi and Sathyamoorthy 2013), chemical bath deposition (Koao et al. 2014), micro-wave assisted sol-gel modification (Alsharaeh et al. 2017), vapour-phase deposition (Haller and Gupta 2014) and laser abolition (Ghorbani 2014), among others. From a broad perspective, wet-chemical co-precipitation is a cost effective and robust method without the need for complex instrumentation for NPs synthesis (Kumbhakar et al. 2016a, Ghosh et al. 2018, Pramanik et al. 2017). Conditional optimization of wet chemical synthesis of pioneered ZnO nanostructures following acidic and basic routes reveals novel secondary (needle topped nano-rod) and tertiary (nano-rosette) nanostructure development (Fig. 1a–f), significant for technological advancement by possible application in devices (Das et al. 2017a). Furthermore, Mishra and Adelung (2017) conducted a review on the synthesis of a novel tetrapod structure of ZnO nanoparticle and its wide application in the field of engineered nanomaterials. ZnO-NPs biosynthesis is also highlighted for crop production (Raliya and Tarafdar 2013).

4. Physico-Chemical Characterization

Specialized instrumentation techniques are required for characteristic realization of nanoscale materials residing between small isolated molecule and bulk aggregates. Distinct physico-chemical attributes, including size, shape, surface property, composition, crystallinity, stability and purity of the engineered nanostructures, are critically investigated following the assessment of optical-near-infra-red absorption tendency (Fig. 2a–c), X-ray and infra-red light scattering ability

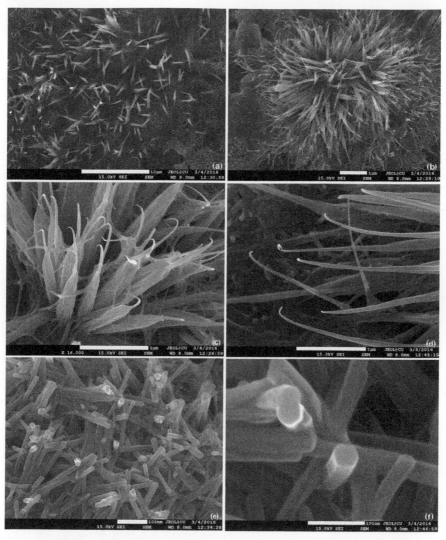

Fig. 1. FESEM micrographs of various ZnO-NS under different reaction environments—monodispersed needle-topped nanorods (a), nanorosettes (b–d) and nanorods (e, f). [Citation: Das et al. 2017a, Nano-Structures and Nano-Objects, 9: 26–30].

(Fig. 2d–h), surface charge potentiality (Fig. 2l–n), etc. (Kumbhakar et al. 2016a, Das et al. 2017a). Individual nanocrystal size is measured either in the form of hydrodynamic diameter under dynamic light scattering analyser (DLS—Fig. 2i–k) or mathematically obtained using Debye-Scherrer equation from X-ray diffraction (XRD) plots. Small angle X-ray scattering (SAXS), XRD and high resolution transmission electron microscope (HR-TEM), coupled with selected area electron diffraction (SAED), are the most efficient techniques to study particle geometry and

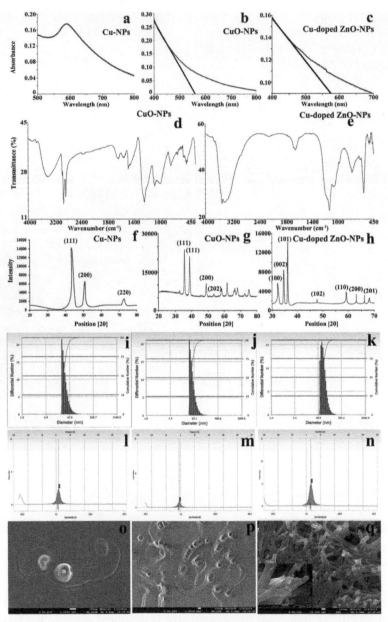

Fig. 2. Opto-physical attributes of nanoparticles (NPs): (a–c) UV-vis absorption plots showing light absorption efficiency in the visible spectrum region, (d–e) FTIR plot showing transmissional depressions at different wavelength regions, (f–h) X-ray diffraction patterns with multiple peak positions at Braggs angles (2θ) demonstrating crystal faces of NPs, (i–k) Bar histograms documenting particle distribution under DLS, (l–n) intensity plots exhibiting surface electric charge potential under Zeta potential analyzer, (o–q) FESEM image showing surface morphology and shape of individual nanocrystal [unpublished].

crystal face morphology, which can be further confirmed visually by field emission scanning electron microscopy (FESEM—Fig. 2o–q) and atomic force microscopy (AFM). Internal architecture along with the lattice fringe of a single nanocrystal can only be studied under HR-TEM.

Chemical composition, bonding nature and purity of NPs are assessed following fourier transform infra-red (FTIR) and mass spectroscopy (MS) and high field (500–600 Hz) nuclear magnetic resonance (NMR). High performance liquid chromatography (HPLC), zeta potency and thermophoresis are techniques that are frequently used for the evaluation of the stability of synthesized nanomaterials in their parent environment.

Complex nanostructures conjugated with biomaterials (specially protein corona) are characterized further using fluorescence correlation spectroscopy (FCS; binding kinetics), tip enhanced raman spectroscopy (TERS; conformational change in protein-NPs conjugate), circular dichroism (CD; dynamic change in structural morphology) and scanning tunnelling microscopy (STM; dispersion-aggregation).

All the techniques employed for opto-physical characterization are used to establish standard nanodimensional quality of the synthesized NPs.

5. Nanoparticle Accumulation and Localization

Plants, being the major component of the ecosystem, are the passage for the uptake and bioaccumulation of nanomaterials in the food chain posing a threat to the host species (Ma et al. 2010, Nair et al. 2010). Nanomaterials are introduced to plants via three basic avenues, seed, root and whole aerial part, either directly or indirectly. Uptake of nanomaterials is analyzed using Atomic Absorption Spectrophotometer (AAS) following tri-acid digestion of NP-treated seedlings (ash) for estimation of the amount of NPs uptaken in ionic form. Uptake of nanomaterials through ICP-MS is also highlighted in watermelon (Raliya et al. 2016).

NPs enter seeds through parenchymatous intercellular spaces, facilitating their diffusion to cotyledons (Van Dongen et al. 2003, Lee et al. 2010, Ma et al. 2010). However, seed treatment experiences the physical barrier nature of seed-coat against NPs uptake (Chichiriccò and Poma 2015), while spraying on leaf documents direct penetration of nanosuspension through stomatal opening and epidermal lesion (Eichert and Goldbach 2008). NP exposure through soil contamination shows varied responsiveness in particle accumulation (Darlington et al. 2009). For such uptake, soil texture, soil-NPs interaction, NPs dissolution in capillary water and root hair permeability, among others, are the major influencing factors (Rico et al. 2011).

Available literatures show that, apart from NPs size and characteristic physico-chemical properties, cell wall composition, mucilage, stomatal aperture, root hair, leaf repellence, membrane porosity, xylem segmentation, casperian strips, hydathodes, lenticel and trichomes are important factors affecting NPs intake and subsequent transportation (Schwab et al. 2015). Internalized NPs by root cell surface invade the vascular cylinder via the primary lateral root junction, where rudimentary casperian strips, along with rapidly growing divisional root tip cells, provide passage for NPs penetration (Lv et al. 2015).

Nanocrystal fractions, irrespective of their mode of penetration, utilize both water (xylem element) and food (phloem elements) conducting systems of the host for their potential distribution in the plant system. For the distribution of such NPs, both root to leaf (López-Moreno et al. 2010) and leaf to root (Hong et al. 2014, Laruea et al. 2014) vascular transportation pathways are involved. In mature plants, casperian strips with high lignification make large voids in the xylem cylinder, providing passages for NPs localization and their uptake along the path of water transport (Cifuentes et al. 2010, Sun et al. 2014, Lv et al. 2015). Apart from vertical distribution of NPs, their corticular transport demonstrates superficial movement through the peripheral network. Vascular transportation increases area of reactivity of NPs by enabling them to reach every corner of the host system. Report shows that not only small sized but also larger NPs are uptaken by roots and transferred to aerial parts (Zhu et al. 2008, Corredor et al. 2009, Hischemoller et al. 2009). A plant cell is surrounded by a rigid cell wall with pores (diameter 3–8 nm; Carpita and Gibeaut 1993) that allow NPs (size smaller than pores of cell wall) to pass through and reach the plasma membrane. Roberts and Oparka (2003) suggested that NPs are transported between the cells via plasmodesmata.

A host cell can take up penetrated NPs either through various endocytic processes, like phagocytosis and pinocytosis (clathrin mediated endocytosis, caveolae mediated endocytosis, clatrin/cavelae independent endocytosis and micropinocytosis), or by ballistic diffusion (Geiser et al. 2005, Limbach et al. 2005, Rothen-Rutishauser et al. 2006). NPs internalization and localization within a plant cell are visualized by confocal laser scanning microscope using fluorescent dyes (hoechst 33342, acridine orange among others). Cellular uptake of NPs is mostly dependent on nanocrystal morphology, geometry and surface reactivity, which act as direct contributing factors for plasma membrane—NPs interaction. However, apart from direct physical internalization, lipid peroxidation and subsequent membrane damage due to NP-mediated stress constitutes a membrane permeable hotspot for uninterrupted nanocrystal entry within the cellular environment. Among the various sub-cellular depositions, NPs localization in nucleus and mitochondria exhibit the most significant impact on cellular health. SEM analysis reveals particle agglomerization of NPs as a part of defence by the host cell. Transmission electron microscope (TEM) facilitates monitoring of NPs distribution in plant tissues at the subcellular level.

6. Physical Attributes

6.1 Seed Germination and Seedling Growth

Assessment of seed germination and seedling growth following NPs treatment provides a preliminary idea on plant growth and development of host. Reports reveal that engineered nanoparticles can induce both positive (Lin et al. 2004, Hojjat and Hojjat 2015) as well as negative (Raskar and Laware 2014, Kumbhakar et al. 2016a) effects on seed germination and seedling growth of higher plants. Inhibitory as well as promotive action of NPs on the said physiological attributes are attributed to particle size (Masarovičová and Kráľová 2013), concentrations (Lee et al. 2010, Elizabath et al. 2017), water and nutrient uptake (Dehkourdi and Mosavi 2013), genotype sensitivity (Ma et al. 2010), etc.

Bright field stereo-zoom and scanning electron microscopy of NP-treated seedlings in relation to controls (Fig. 3a–c, 4a) exhibit variable morphological abnormalities, such as radicle bifurcation (Fig. 4b), root cap deformation (Fig. 3g–i), root hair suppression (Fig. 3i), radicle necrosis (Fig. 3d–e), plumule browning and necrosis (Fig. 3f), leaf epidermal rupturing (Fig. 4e–f) and leaf surface lesion (Fig. 4c–d), among others. Such abnormalities highlight nano-bio interaction at the interface level.

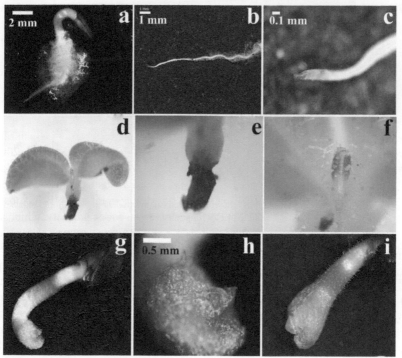

Fig. 3. Stereomicroscopic images showing morphological deformities in seedlings induced by NPs (d–i) compared to controls (a–c) [unpublished].

7. Oxidative Stress Induction and Antioxidant Enzyme Responses

7.1 Root Cell Viability

Reactive oxygen species (ROS) generation due to NP exposure in roots of various plant species are indicated by excess Evans blue uptake, resulting in cytotoxicity

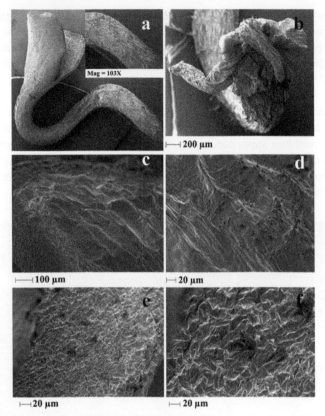

Fig. 4. Scanning electron micrographs showing structural anomalies in seedlings induced by nanoparticles [unpublished].

(Pan et al. 2001, Nair and Chung 2015). Evans blue dye specifically binds to damaged cell membrane and determines the extent of the damage caused by variable doses of NPs. Passardi et al. (2005) reported that a plant under stress stimulates cell wall lignification, resulting in reduction of root cell viability along with root growth inhibition.

7.2 Oxidative Stress

Direct and indirect ROS production primarily depends on physico-chemical properties of NPs and the concerned host genotype. ROS are the by-product of aerobic metabolism, comprising of free radicals like hydroxyl ($\cdot$OH) and superoxide (O^{2-}) species, singlet oxygen ($\cdot O_2$) (Gill and Tuteja 2011), among others, result in oxidative stress induction. Such energized super radicals are later converted to hydrogen peroxide (H_2O_2) by the scavenging action of different classes of superoxide dismutase and catalase (Imlay 2008). Oxidative stress is considered as a determinant marker of NP-mediated phytotoxicity. Elevated ROS due to NP exposure surpasses the defence mechanism in the plant system, inducing DNA damage, lipid peroxidation, membrane damage, protein oxidation, etc. (Meriga et al. 2004, Sharma et al. 2012, Tripathy and Oelmüller 2012), causing cell death. NPs internalization in the cellular environment due to nano-bio interaction results in disruption of cell signalling and redox homeostasis (Panda et al. 2017).

NP-treated cells generate oxidative stress, showing an increase in concentration of malondialdehyde (MDA). MDA is the outcome of enhanced decomposition of polyunsaturated fatty acids in membranes as an indicator of lipid peroxidation conferring membrane damage (Mittler 2002, Tanou et al. 2009, Halliwell and Gutteridge 2015). H_2O_2 (primary indicator of stress accumulation due to its enhanced half-life) and MDA show differential elevation (Majumdar et al. 2014, Yang et al. 2017) in different plant species following NPs treatments, suggesting asynchrony in host response to NP-induced oxidative stress.

7.3 Antioxidant Responses

Nanomaterials inducing stress in cellular system of plant species triggers defence response in the form of enzymatic and non-enzymatic antioxidants. Anti-oxidants enzymes, like superoxide dismutase (SOD), catalase (CAT), ascorbate peroxidase (APX), guaiacol peroxide (GPX), glutathione reductase (GR), glutathione-S-transferase (GST), monodehydroascorbate reductase (MDAR) and dehydroascorbate reductase (DHAR), and non-enzymatic antioxidants, namely, ascorbic acid (AA) and reduced glutathione (GSH), are reported to play a crucial role in combatting NP-mediated stress accumulation in host and to attain redox equilibrium (Dimkpa et al. 2012, Das et al. 2017b, Tripathi et al. 2017). However, NPs types and uptake, antioxidant responses, concentration/doses employed, duration of treatment and genotype sensitivity are the significant contributing factors for proper balancing of oxidative stress and defence.

Increase in APX (EC 1.11.1.11) reaction kinetics in NP-challenged seedlings demonstrates oxidative defence response of the plant system to maintain a favorable redox state (Caverzan et al. 2012) by constructing a reduced ascorbate pool (Smirnoff 2011). H_2O_2 scavenging activity of APX results in production of monodehydroascorbate (MDA), which is reduced to ascorbate by MDAR using NADPH as the specific electron donor (Chakradhar et al. 2017). Co-ordinated functioning of APX-MDAR possibly results in effective recycling of ascorbate pool favorable for redox homeostasis. SOD (EC 1.15.1.1), a metalloenzyme, acts as second line of defence in host species and is produced against oxidative stress and catalyzes dismutation of highly toxic free radical $^{\cdot}O^{2-}$ to less toxic H_2O_2 and O_2 (Scandalios 1993) as well as preventing $^{\cdot}OH$ radical formation by Haber-Weiss reaction (Das and Roychoudhary 2014). Elevation in cellular SOD concentration enhances the stress tolerant potentiality of host species (Rajeshwari et al. 2015, Ma et al. 2016, Zhang et al. 2017). GR (EC 1.8.5.1) is an NADPH-dependent antioxidant enzyme, contributing to both enzymatic and non-enzymatic oxidation-reduction cycles. The enzyme is responsible for maintenance of the GSH/GSSG ratio in the cellular compartment, following oxidation of GSH to GSSG (Gossett et al. 1996, Rao et al. 2006, Yang et al. 2017). Zagorchev et al. (2013) opined that elevation in GST (EC 2.5.1.18) enzyme kinetics in relation to NPs treatment promotes the thiol-dependent ROS scavenging system, thereby minimizing peroxidase toxicity. APX-MDAR-GR-GST synchrony is significant for defence reactions against oxidative stress in host cellular systems. GSH coupled with GPX also minimizes lipid peroxidation (Thomas et al. 1990).

Catalase (EC 1.11.1.16), a tetrameric heme-containing enzyme, possesses strong affinity towards H_2O_2 and converts it into water and oxygen, thereby minimizing stress accumulation. The enzyme is present in cytosol, chloroplast, mitochondria and peroxisomes, and requires a smaller amount of energy for catabolic reaction (Das and Roychoudhary 2014). Elevated catalase enzyme is observed in NP-treated root, stem and leaves (Hernandez-Viezcas et al. 2011, Rao and Shekhawat 2016). However, the enzyme activation is non-synchronous and mostly exhibits moderate reduction in catabolic activity at higher concentration of NPs treatments (Siddiqi and Husen 2017). NPs agglomeration at high concentrations (Siddiqi and Husen 2017) and prolonged duration results in lower stress responsiveness (reduced NPs accumulation in biological system) contributing to suppression in catalase activity.

Although NPs treatment induces oxidative stress and triggers antioxidant enzyme responses in the host, persistence of the cellular stress environment prevails due to asynchrony in the defence enzyme kinetics.

7.4 Apoptotic Cell Death

Elevated ROS generation due to phyto-nano interaction causes cell cycle dysregulation, resulting in cell granularity and shrinkage of the cell at an early phase of apoptosis (Vamanu et al. 2008, Faisal et al. 2013) leading to enhanced cell mortality. Cell cycle checkpoints (the G_1/S checkpoint and the G_2/M checkpoint) are used to examine and regulate cell cycle progression, which is either altered or inhibited in NPs treatments. Mahmoudi et al. (2011) revealed that apoptotic cells due to NP exposure may be the outcome of failure in separation of condensed chromosomes in mitotic progression. Flow cytometric analysis confirmed that NPs are capable of inducing sub-G_1 apoptotic peak (different amounts of DNA lost in form of fragments) or G_2/M arrest (Chunyan and Valiyaveettil 2013) in host seedlings in relation to controls (Fig. 5). Such events are activated by the caspase-3 protease pathway of signal transduction (Green 1998, Eom and Choi 2010, Li and Xing 2011, Faisal et al. 2013).

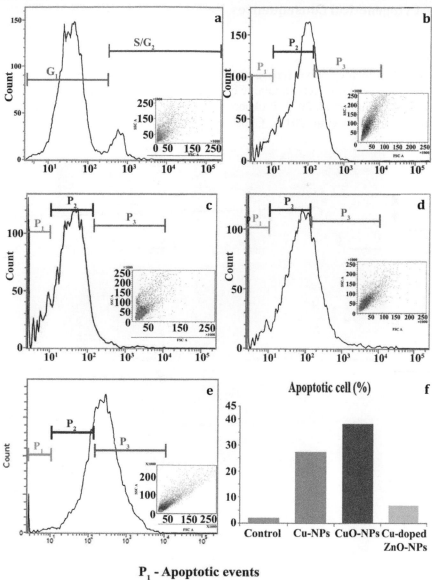

P₁ - Apoptotic events

P₂ - G₁ phase distribution

P₃ - S/G2 phase distribution

Fig. 5. Assessment of cell cycle distribution pattern of root tip nuclei in control (a) and in NPs treatments (b-Cu-NPs; c,d-CuO-/NPs; e-ZnO-NPs), following flow cytometric analysis. (f) Bar histogram representing apoptotic cell death in NP-treated root tip cells [unpublished].

8. Cytogenetical Consequences

8.1 Mitotic Aberrations

Different classes of NPs, like Ag, TiO_2, ZnO, CuO, Cu and CdS, among others, are assessed for their cytotoxic (enhancement in chromosomal aberration frequency) and mitodepressive effects on mitotic cells of various plant species (Patlolla et al. 2012, Shaymurat et al. 2012, Prokhorova et al. 2013, Abou-Zeid and Moustafa 2014, Kumbhakar et al. 2016a, Ghosh et al. 2017a) in order to understand the

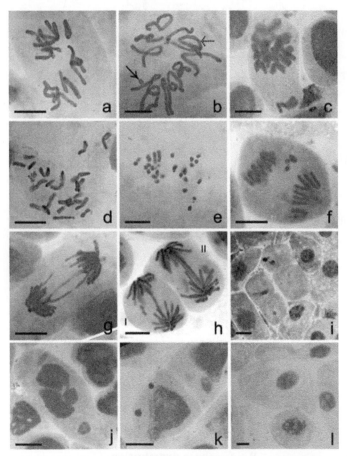

Fig. 6. (a–l) Mitotic cells of *N. sativa* in (a) control and (b-l) Cu- and CdS-NPs treated material. (a) 2n = 12, (b) differentially condensed chromosome with a fragment (right-handed arrow) and a ring (dotted arrow), (c) diplochromatid nature of chromosomes with unoriented chromosomes, (d) cell with > 2n = 12 chromosomes, (e) differential condensation and fragmentation of chromosomes, (f) anaphase with lagging fragments, (g) chromosome bridge formation, (h) I: anaphasic bridge with two equal size fragments, II: double bridge with two minute fragments, (i) cells without or with depleted chromatin, (j) cell with fragmented chromatin mass, (k) interphase cell with two unequal sized micronuclei, (l) giant cell. Scale bar = 20 μm. [Citation: Kumbhakar et al. 2016a, Journal of Experimental Nanoscience 11: 823–839].

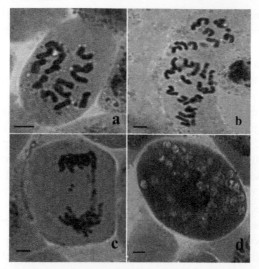

Fig. 7. (a–d) NP-induced mitotic configurations in *Lathyrus sativus* (a) 2n = 14. (b) Polyploidy 2n = 28, (c) Bridge with paired fragments, (d) Vacuolated giant cell. Scale bar = 10 μm. [Citation: Ghosh et al. 2017a, Cytologia 82: 267–271].

effects of nanomaterials at a chromosomal level. NP-induced reduction in mitotic index (MI) may be the outcome of S-phase blockage in the cell cycle (Kim et al. 2012). The enhanced cytological abnormalities compared to normal cell (stickiness, ring and bridge formation, fragmentation including localized breakage—Fig. 7c, diplochromatid nature, laggards and polyploid cell and micronuclei formation —Fig. 6a–l, Fig. 7a–b) are the consequences of mitotic spindle impairment and chromosomal alteration at various stages of cell cycle (Kumbhakar et al. 2016a, Ghosh et al. 2017a, Pramanik et al. 2018). Cell and nuclear shape deformation, irregular cell plate formation and occurrence of giant cells suggest abnormalities in the cellular metabolism induced by NP exposure (Kumbhakar et al. 2016a, Ghosh et al. 2017a, Pramanik et al. 2018). However, intense vacuolation (Fig. 7d) noted in NP-treated root cells is indicative of minimization potentiality of the host species against nanomaterials internalization in the cellular system (Ghosh et al. 2017a). Comparative assessment of mitotic aberrations following the use of CdS-NPs show differential responses in *N. sativa* (Kumbhakar et al. 2016a), *Lathyrus sativus* (Ghosh et al. 2017a) and *Coriandrum sativum* (Pramanik et al. 2018). *N. sativa* seems to be affected more than the other two germplasms and it may be due to chromosome length, chromosome number, heterochromatin content, moisture content, seed coat barrier, etc.

8.2 Meiotic Chromosomal Anomalies

Several mitotic cycles lead to meiotic cell divisions. Available evidences till date indicate that rare studies are performed on the effects of NPs on meiotic cells in plant species (Kumbhakar et al. 2016a, Saha and Dutta Gupta 2017). Such studies are potentially important for genetic consequences, as the meiotic outcome will be transmitted through the gamete in the following generations. Cu- and CdS-NPs are reported to induce meiotic aberrations, namely, sticky chromosomes, fragmentation, lack of pairing between homologues, differential chromosome condensation, aneuploid variation, metaphase I grouping, laggards, unequal separation, bridges at anaphase I and anaphase II, tripolarity, unequal multipolar groups and disorganised groups of chromosomes in *Nigella sativa* L. (Kumbhakar et al. 2016a). Such aberrations induced by NP exposure and compared with that of conventional mutagens (ethyl methanesulphonate and gamma irradiations) reflect more similarity with gamma irradiations. However, the potential energy provided by NPs to cause such intense chromosomal fragmentations is rather difficult to ascertain and needs extensive exploration.

Chromosome aberrations in the meiotic cells in relation to normal (Fig. 8a–x) yields microgametes with reduced viability, which disrupts vital processes of plant reproduction, highlighting the possibility of genetic changes that can bring about phenotypic alterations in plant ideotype in subsequent generations (Kumbhakar et al. 2016b, Ghosh et al. 2017b, Saha and Dutta Gupta 2017). NPs affecting both mitotic and meiotic cells of plants might affect the signal transduction pathway controlling the different phases of plant cell division. Cytotoxic effects of nanomaterials in plant systems must, therefore, be given more emphasis in future studies for proper understanding of NPs interaction with chromosomes and that with cellular system.

9. DNA Damage

NPs induce genotoxicity (Landsiedel et al. 2009, Kovacic and Somanathan 2010, Phugare et al. 2011, Atha et al. 2012, de Lima et al. 2012) by interacting with plant DNA. Nanomaterials generate $^{\cdot}OH$ radical in the cellular system, which is capable of inducing DNA strand breakage (Faisal et al. 2016). The available literature suggests that oxidative stress due to NP exposure causes ionic imbalance, resulting in oxidation of purine molecules (Phugare et al. 2011, Pakrashi et al. 2014), causing DNA damage (Xi et al. 2004). DNA damage is either quantitatively assessed following the use of comet assay (measurement of tail DNA—Fig. 9, Pourrut et al. 2015, Ghosh et al. 2018) or qualitatively measured from DNA smearing under laddering assay (Kasibhatla et al. 2006, Chung et al. 2015). DNA damage results in cell cycle arrest and apoptosis, resulting in genetic instability (Singh et al. 2009). Ghosh et al. (2012) reported that the extent of DNA damage in NP-treated roots is higher than leaves, as NPs are directly up taken by roots and later transported to leaves. Further study on the interpretation of gene expression is required for better understanding of DNA damage machinery caused by NPs internalization in plants. Various factors are reported to be responsible for NP-mediated genotoxicity, such as concentration, particle size, high surface to volume ratio, physicochemical properties, stability, aggregation, experimental conditions (temperature and time) and method of exposure (seeds/seedlings/mature

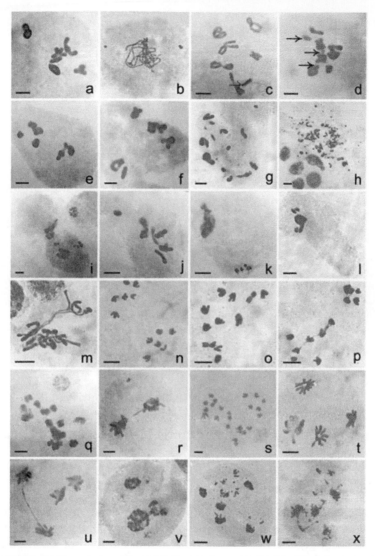

Fig. 8. Meiosis in (a) control and (b–x) NP-treated plants. (a) 6II, (b) pachytene stage with two fragments, (c) diakinesis cell showing breakage of an arm (dotted arrow) in a bivalent, (d) MI with 5II + 2I + 3 fragments (right-handed arrow), (e–f) chromosomal grouping (4II + 2II) at MI, (g) fragments in association to bivalent and univalent at MI possibly 2n > 12, (h) intense chromosomal fragmentation at MI, (i) cytomictic behaviour involving two meiocytes, (j) hyperploid PMC—7II at MI, (k) hypoploid cell with condensed type chromosomal fragments, (l) agglutination of chromatin and fragments, (m) differential condensation of chromosomes at MI, (n) 6/6 separation at AI, (o) 3/5/4 separation of chromosomes at AI, (p) 5/7 AI separation with interchromosomal connections, (q) 2n > 12 with sticky chromosomes at AI, (r) dicentric chromatid bridge with an acentric fragment, (s) 2n = 24; AI showing tripolar organisation of chromosomes, (t) AII with one disorganised pole, (u) AII with bridge, (v) unequal tripolarity, (w, x) AII with multiple groups and fragments. Scale bar = 10 μm.
[Citation: Kumbhakar et al. 2016a, Journal of Experimental Nanoscience 11: 823–839].

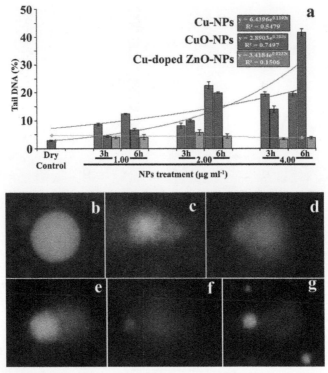

Fig. 9. Comet photomicrographs (% tail DNA) in (a) control and in (b–g) NPs treatments [unpublished].

plants) (Gao et al. 2006, Liu et al. 2010, Kim et al. 2011, Rico et al. 2011, de Lima et al. 2012, Gaiser et al. 2012).

10. Heritable Consequences

Kumbhakar et al. (2016a) reported that Cu- and CdS-NPs can induce chromosomal alteration in meiotic cells of *N. sativa* and are of the opinion that such aberrations can bring about genetic variations in progenies in the form of macromutations (Kumbhakar et al. 2016b). Singh et al. (2009), from Comet assay and micronuclei testing, suggested mutagenic potentiality of Ag-NPs. DNA lesions are reported to be caused by NPs which mispair with normal bases during the process of replication that can bring about mutagenic changes (Vidal et al. 2001). Nel et al. (2006) reported that carbon nanotubes (CNTs) interacting with DNA causes abnormal effects, DNA damages or mutagenesis. Langie et al. (2015) highlighted that alkaline comet assay is an indicator test for assessing intermediate DNA lesions that can be repaired and fixed into mutations. Vecchio et al. (2011) first reported phenotypic alterations in *Drosophila melanogaster*, following Au-NPs treatments; moreover, they found that such alterations were also transmitted to the descendants. In plant

Fig. 10. Plant types in *Macrotyloma uniflorum*—(a) control plant, *(b) bushy, (c) prostrate, (d)* seed coat color of normal (i) and mutant (ii). [Citation: Halder et al. 2015a, Genetic Resources and Crop Evolution 62: 165–175].

Fig. 11. Normal (a) and mutant (b–d) plant types of *Sesamum indicum*. (a) Normal, (b) *quadra-fruits per node*, (c) *tri-fruits per node*, (d) *multilocular fruits*. [Citation: Das et al. 2017, Indo American Journal of Pharmaceutical Sciences 4: 3815–3820].

species (*Macrotyloma uniflorum*), Halder et al. (2015a,b) pioneerly documented NP-induced, stable, heritable phenotypic alterations (macromutations—Fig. 10a–d). NP-induced macromutations are also reported in *Nigella sativa* (Kumbhakar et al. 2016b), *Sesamum indicum* (Fig. 11a–d, Das et al. 2017b) and in *Indigofera tinctoria* (Ghosh et al. 2017b). The raised phenotypic mutants either bred true at M_3 or they segregated at F_2 in accordance with the Mendelian pattern of segregation. The macromutant types raised, following NP treatment, are found to be comparable with those of conventional mutagens, like gamma irradiations and ethyl methanesulphonate. However, extensive studies are still required in different plant species before considering NPs as an alternative source of mutagens. If so, NPs will open up new vista in agricultural science in relation to crop improvement.

Conclusion

From the above discussion on phyto-NP interaction, some key questions arise: (1) Is there any specific signal transduction pathway that exists for NP uptake in the cellular system and is it energy dependent or not? (2) Since the host always attempts to minimize the effect of NPs by converting them to either an agglomerated form or breaking them into ionic species, what is the mechanism that successfully maintains nanomaterials in an effective form in plant systems? (3) Localization of NPs is reported in both the cytoplasm and the nucleus, so what are the contributing effects of the nanomaterials that are imparted on cellular compartments other than nucleus? (4) Genotype sensitivity is the determinant factor in phyto-NPs interaction, so is there any differential mechanism existing for bio-effectivity? (5) How is the genomic profile altered in the host species as a consequence of NPs interacting with DNA binding motifs? (6) Chromosome data highlights the role of NPs in inducing chromosomal breaks and subsequent abnormalities alike to irradiations, so from where do NPs attain energy? (7) What is the ultimate fate of NPs in the host cellular environment? Unravelling of many such questions through extensive research can open a new dimension in nanoscience.

11. Future Prospects

Investigation involving phyto-NPs interactions so far is rather limited in rooted plant species. Genotype responses against NP exposure need to be studied specifically using species of commercial interest and food values. Study associated with nanomaterial penetration to plant system requires novel confirmatory technique(s) in order to identify plant specific gateways of NP entry. Although extensive research is carried out to understand intracellular localization of NPs, their precise movement within the cytoskeleton embedded cytoplasmic matrix is still to be revealed. Cytotoxic as well as genotoxic potentiality is well known for nanodimensional class materials, but understanding their effectivity on plant genomic, proteomic and metabolic profiling is limited. Possible plant specific repair mechanism(s) functioning under DNA damage need to be investigated in order to assess the restoration attempts of the host genome profile under a NP-stressed environment. DNA microarray along with real time PCR analyses can provide valuable insight towards understanding the dynamic alteration of plant genomic profile through direct and indirect interaction with NPs. Liquid chromatography coupled with tandem mass spectrometric (LC-MS/MS) approach can also highlight deviation in proteomics, as well as metabolic profile, under NP-penetrated cytoplasm. Therefore, studies should be focused on experimental novelty in order to understand phyto-NPs interaction even at nano-dimensional resolution.

Acknowledgement

The authors are thankful to UGC DAE consortium (Kolkata) and CRNN, Calcutta University, for technical help.

References

Abou-Zeid, H.M. and Y. Moustafa. 2014. Physiological and cytogenetic responses of wheat and barley to silver nanopriming treatment. Int. J. Appl. Biol. Pharm. 5: 265–278.

Ali, A., H. Zafar, M. Zia, I. ulHaq, R.A. Phull, J.S. Ali and A. Hussain. 2016. Synthesis, characterization, applications, and challenges of iron oxide nanoparticles. Nanotechnol. Sci. Appl. 9: 49–67.

Alsharaeh, E.H., T. Bora, A. Soliman, F. Ahmed, G. Bharath, M.G. Ghoniem, K.M. Abu-Salah and J. Dutta. 2017. Sol-gel-assisted microwave-derived synthesis of anatase Ag/TiO$_2$/GO Nanohybrids toward efficient visible light phenol degradation. Catalysts 7: 133.

Anjum, M., R. Miandad, M. Waqas, F. Gehany and M.A. Barakat. 2016. Remediation of wastewater using various nano-materials. Arab. J. Chem. http://dx.doi.org/10.1016/j.arabjc.2016.10.004.

Atha, D.H., H. Wang, E.J. Petersen, D. Cleveland, R.D. Holbrook, P. Jaruga, M. Dizdaroglu, B. Xing and B.C. Nelson. 2012. Copper oxide nanoparticle mediated DNA damage in terrestrial plant models. Environ. Sci. Technol. 46: 1819–1827.

Bahadar, H., F. Maqbool, K. Niaz and M. Abdollahi. 2016. Toxicity of nanoparticles and an overview of current experimental models. Iran. Biomed. J. 20: 1–11.

Benckiser, G. 2017. Nanotechnology in life science: its application and risk. *In*: Prasad, R., M. Kumar and V. Kumar (eds.). Nanotechnology. Springer, Singapore. https://doi.org/10.1007/978-981-10-4573-8_2.

Bhatia, S. 2016. Nanoparticles Types, Classification, Characterization, fabrication methods and drug delivery applications. Cham. pp. 33 – 93. *In:* Natural Polymer Drug Delivery Systems. Springer. https://doi.org/10.1007/978-3-319-41129-3_2.

Biffi, S., R. Voltan, E. Rampazzo, L. Prodi, G. Zauli and P. Secchiero. 2015. Applications of nanoparticles in cancer medicine and beyond: Optical and multimodal *in vivo* imaging, tissue targeting and drug delivery. Expert Opin. Drug Deliv. 12: 1837–1849.

Boal, A.K., K. Das, M. Gray and V.M. Rotello. 2002. Monolayer Exchange Chemistry of γ-Fe$_2$O$_3$ Nanoparticles. Chem. Mater. 14: 2628–2636.

Carpita, N.C and D.M. Gibeaut. 1993. Structural models of primary cell walls in flowering plants: Consistency of molecular structure with the physical properties of the walls during growth. Plant J. 3: 1–30.

Caverzan, A., G. Passaia, S.B. Rosa, C.W. Ribeiro, F. Lazzarotto and M. Margis-Pinheiro. 2012. Plant responses to stresses: role of ascorbate peroxidase in the antioxidant protection. Genet. Mol. Biol. 35: 1011–1019.

Chakradhar, T., S. Mahanty, R.A. Reddy, K. Divya, P.S. Reddy and M.K. Reddy. 2017. Reactive Oxygen Species and Antioxidant Systems in Plants: Role and Regulation under Abiotic Stress. Springer Nature, Singapore.

Chichiriccò, G. and A. Poma. 2015. Penetration and toxicity of nanomaterials in higher plants. Nanomaterials (Basel) 5: 851–873.

Chung, I.-M., A.A. Rahuman, S. Marimuthu, A.V. Kirthi, K. Anbarasan and G. Rajakumar. 2015. An investigation of the cytotoxicity and caspase-mediated apoptotic effect of green synthesized zinc oxide nanoparticles using *Eclipta prostrata* on human liver carcinoma cells. Nanomaterials 5: 1317–1330.

Chunyan, W. and S. Valiyaveettil. 2013. Correlation of biocapping agents with cytotoxic effects of silver nanoparticles on human tumor cells. RSC Adv. 3: 14329–14338.

Cifuentes, Z., L. Custardoy, J.M. de la Fuente, C. Marquina, M.R. Ibarra, D. Rubiales and A. Pérez-de-Luque. 2010. Absorption and translocation to the aerial part of magnetic carbon-coated nanoparticles through the root of different crop plants. J. Nanobiotechnology 8: 26.

Corredor, E., P.S. Testillano, M.J. Coronado, P. González-Melendi, R. Fernández-Pacheco, C. Marquina, M.R. Ibarra, J.M. de la Fuente, D. Rubiales, A. Pérez-de-Luque and M.-C. Risueño. 2009. Nanoparticle penetration and transport in living pumpkin plants: *In situ* subcellular identification. BMC Plant Biol. 9: 45.

Darlington, T.K., A. Neigh, M. Spencer, O. Nguyen and A. Oldenburg. 2009. Nanoparticle characteristics affecting environmental fate and transport through soil. Environ. Toxicol. Chem. 28: 1191–1199.

Das, D., A.K. Datta, D.V. Kumbhakar, B. Ghosh, A. Pramanik and S. Gupta. 2017a. Conditional optimisation of wet chemical synthesis for pioneered ZnO nanostructures. Nano-Structures & Nano-Objects 9: 26–30.

Das, D., A.K. Datta, D.V. Kumbhakar, B. Ghosh and A. Pramanik. 2017b. Assessment of antioxidant enzyme responses in host (*Sesamum indicum* L. Pedaliaceae) following nanoparticles (Copper, Copper Oxide and Copper Doped Zinc Oxide) exposure. Int. J. Curr. Adv. Res. 6: 5512–5516.

Das, K and A. Roychoudhury. 2014. Reactive oxygen species (ROS) and response of antioxidants as ROS-scavengers during environmental stress in plants. Front. Environ. Sci. 2: 53.

de Lima, R., A.B. Seabra and N. Durán. 2012. Silver nanoparticles: A brief review of cytotoxicity and genotoxicity of chemically and biogenically synthesized nanoparticles. J. Appl. Toxicol. 32: 867–879.

Dehkourdi, E.H. and M. Mosavi. 2013. Effect of anatase nanoparticles (TiO_2) on parsley seed germination (*Petroselinum crispum*) *in vitro*. Biol. Trace Elem. Res. 155: 283–286.

Di Bucchianico, S., F. Cappellini, F. Le Bihanic, Y. Zhang, K. Dreij and H.L. Karlsson. 2017. Genotoxicity of TiO_2 nanoparticles assessed by mini-gel comet assay and micronucleus scoring with flow cytometry. Mutagenesis 32: 127–137.

Dimkpa, C., J. Mclean, D. Latta, E. Manangón, D. Britt, W. Johnson, M.I. Boyanov and A.J. Anderson. 2012. CuO and ZnO nanoparticles: Phytotoxicity, metal speciation, and induction of oxidative stress in sand-grown wheat. J. Nanopart. Res. 14: 1–15.

Dumala, N., B. Mangalampalli, S. Chinde, S.I. Kumari, M. Mahoob, M.F. Rahman and P. Grover. 2017. Genotoxicity study of nickel oxide nanoparticles in female Wistar rats after acute oral exposure. Mutagenesis 32: 417–427.

Eichert, T. and H.E. Goldbach. 2008. Equivalent pore radii of hydrophilic foliar uptake routes in stomatous and astomatous leaf surfaces–further evidence for a stomatal pathway. Physiol. Plant 132: 491–502.

Elizabath, A., V. Bahadur, P. Mistra, V.M. Prasad and T. Thomas. 2017. Effect of different concentrations of iron oxide and zinc oxide nanoparticles on growth and yield of carrot (*Daucus carota* L.). J. Pharmacogn. Phytochem. 6: 1266–1269.

Eom, H.-J. and J. Choi. 2010. p38 MAPK activation, DNA damage, cell cycle arrest and apoptosis as mechanisms of toxicity of silver nanoparticles in Jurkat T cells. Environ. Sci. Technol. 44: 8337–8342.

Faraji, M., Y. Yamini and M. Rezaee. 2010. Magnetic nanoparticles: Synthesis, stabilization, functionalization, characterization, and applications. J. Iran. Chem. Soc. 7: 1–37.

Faisal, M., Q. Saquib, A.A. Alatar, A.A. Al-Khedhairy, A.K. Hegazy and J. Musarrat. 2013. Phytotoxic hazards of NiO-nanoparticles in tomato: A study on mechanism of cell death. J. Hazard. Mater. 250-251: 318–332.

Faisal, M., Q. Saquib, A.A. Alatar, A.A. Al-Khedhairy, M. Ahmed, S.M. Ansari, H.A. Alwathnani, S. Dwivedi, J. Musarrat and S. Praveen. 2016. Cobalt oxide nanoparticles aggravate DNA damage and cell death in eggplant via mitochondrial swelling and NO signaling pathway. Biol. Res. 49: 20.

Frank, S., P. Poncharal, Z.L. Wang and W.A. De Heer. 1998. Carbon nanotube quantum resistors. Science 280: 1744–1746.

Frankamp, B.L., A.K. Boal, M.T. Tuominen and V.M. Rotello. 2005. Direct control of the magnetic interaction between iron oxide nanoparticles through dendrimer-mediated self-assembly. J. Am. Chem. Soc. 127: 9731–9735.

Gaiser, B.K., T.F. Fernandes, M.A. Jepson, J.R. Lead, C.H. Tyler, M. Baalousha, A. Biswas, G.J. Britton, P.A., Coles, B.D. Johnston, Y. Ju-Nam, P. Rosenkranz, T.M. Scown and V. Stone. 2012. Interspecies comparisons on the uptake and toxicity of silver and cerium dioxide nanoparticles. Environ. Toxicol. Chem. 31: 144–154.

Gao, F.Q., F.S. Hong, C. Liu, L. Zheng and M.Y. Su. 2006. Mechanism of nano-anatase TiO_2 on promoting photosynthetic carbon reaction of spinach: Inducing complex of Rubisco–Rubisco activase. Biol. Trace Elem. Res. 111: 286–301.

Geiser, M., B. Rothen-Rutishauser, N. Kapp, S. Schurch, W. Kreyling, H. Schulz, M. Semmler, V. Im Hof, J. Heyder and P. Gehr. 2005. Ultrafine particles cross cellular membranes by nonphagocytic mechanisms in lungs and in cultured cells. Environ. Health Perspect. 113: 1555–1560.

Georgakilas, V., J.A. Perman, J. Tucek and R. Zboril. 2015. Broad family of carbon nanoallotropes: Classification, chemistry, and applications of fullerenes, carbon dots, nanotubes, graphene, nanodiamonds, and combined superstructures. Chem. Rev. 115: 4744–4822.

Ghorbani H.R. 2014. A review of methods for synthesis of Al nanoparticles. Orient. J. Chem. 30: 1941–1949.

Ghosh, M., J.M., S. Sinha, A. Chakraborty, S.K. Mallick, M. Bandyopadhyay and A. Mukherjee. 2012. *In vitro* and *in vivo* genotoxicity of silver nanoparticles. Mutat. Res. 749: 60–69.

Ghosh, B., A.K. Datta, A. Pramanik, D.V. Kumbhakar, D. Das, R. Paul and J. Biswas. 2017a. Mutagenic effectivity of cadmium sulphide and copper oxide nanoparticles on some physiological and cytological attributes of *Lathyrus sativus* L. Cytologia 82: 267–271.

Ghosh, B., A.K. Datta, D. Das, D.V. Kumbhakar and A. Pramanik. 2017b. Nanoparticles mediated phenotypic mutation in *Indigofera tinctoria* L. (family: Fabaceae). Int. J. Res. Ayurveda Pharm. 8: 290–295.

Ghosh, B., A.K. Datta, D. Das, D.V. Kumbhakar and A. Pramanik. 2018. Assessment of nanoparticles (copper, cadmium sulphide, copper oxide and zinc oxide) mediated toxicity in a plant system (*Indigofera tinctoria* L.; Fabaceae). Res. J. Chem. Environ. (In Press).

Gill, S.S. and N. Tuteja. 2011. Cadmium stress tolerance in crop plants: Probing the role of sulfur. Plant Signal Behav. 6: 215–222.

Gossett, D.R., S.W. Banks, E.P. Millhollon and M.C. Lucas. 1996. Antioxidant response to NaCl-stress in a control and an NaCl-tolerant cotton cell line grown in the presence of paraquat, buthioninsulfoximine, and exogenous glutathione. Plant Physiol. 112: 803–806.

Green, D.R. 1998. Apoptotic pathways: The roads to ruin. Cell 94: 695–698.

Halder, S., A. Mandal, D. Das, S. Gupta, A.P. Chattopadhyay and A.K. Datta. 2015a. Copper nanoparticle induced macromutation in *Macrotyloma uniflorum* (Lam.) Verdc. (Family: Leguminosae): a pioneer report. Genet. Resour. Crop Evol. 62: 165–175.

Halder, S., A. Mandal, D. Das, A.K. Datta, A.P. Chattopadhyay, S. Gupta and D.V. Kumbhakar. 2015b. Effective potentiality of synthesised CdS nanoparticles in inducing genetic variation on *Macrotyloma uniflorum* (Lam.) Verdc. BioNanoSci. 5: 171–180.

Haller, P.D. and M. Gupta. 2014. Synthesis of polymer nanoparticles via vapor phase deposition onto liquid substrates. Macromol. Rapid Commun. 35: 2000–2004.

Halliwell, B. and J.M. Gutteridge. 2015. Free radicals in biology and medicine. USA: Oxford University Press.

Hao, C., F. Feng, X. Wang, M. Zhou, Y. Zhao, C. Ge and K. Wang. 2015. The preparation of Fe_2O_3 nanoparticles by liquid phase-based ultrasonic-assisted method and its application as enzyme-free sensor for the detection of H_2O_2. RSC Adv. 5: 21161–21169.

Hernandez-Viezcas, J.A., H. Castillo-Michel, A.D. Servin, J.R. Peralta-Videa and J.L. Gardea-Torresdey. 2011. Spectroscopic verification of zinc absorption and distribution in the desert plant *Prosopis juliflora–velutina* (velvet mesquite) treated with ZnO nanoparticles. Chem. Eng. J. 170: 346–352.

Hischemoller, A., J. Nordmann, P. Ptacek, K. Mummenhoff and M. Hasse. 2009. *In vivo* imaging of the uptake of upconversion nanoparticles by plant roots. J. Biomed. Nanotechnol. 5: 278–284.

Hojjat, S.S and H. Hojjat. 2015. Effect of nano silver on seed germination and seedling growth in fenugreek seed. Int. J. Food Eng. 1: 106–110.

Hong, J., J.R. Peralta-Videa, C. Rico, S. Sahi, M.N. Viveros, J. Bartonjo, L. Zhao and J.L. Gardea-Torresdey. 2014. Evidence of translocation and physiological impacts of foliar applied CeO_2 nanoparticles on Cucumber (*Cucumis sativus*) plants. Env. Sci. Technol. 48: 4376–4385.

Imlay, J.A. 2008. Cellular defences against superoxide and hydrogen peroxide. Annu. Rev. Biochem. 77: 755–776.

Jin, X., M. Deng, S. Kaps, X. Zhu, I. Hölken, K. Mess, R. Adelung and Y.K. Mishra. 2014. Study of tetrapodal ZnO-PDMS composites: A comparison of fillers shapes in stiffness and hydrophobicity improvements. PLoS ONE 9: e106991.

Kasibhatla, S., G.P. Amarante-Mendes, D. Finucane, T. Brunner, E. Bossy-Wetzel and D.R. Green. 2006. Analysis of DNA fragmentation using agarose gel electrophoresis. Cold Spring Harb. Protoc. 2006: pdb.prot4429.

Kim, Y.T., J.H. Han, B.H. Hong and Y.U. Kwon. 2010. Electrochemical deposition of CdSe quantum dot arrays on large-scale graphene electrodes using mesoporous silica thin film templates. Adv. Mater. 22: 515–518.

Kim, J.S., J.H. Sung, J.H. Ji, K.S. Song, J.H. Lee, C.S. Kang and I.J. Yu. 2011. *In vivo* genotoxicity of silver nanoparticles after 90-day silver nanoparticle inhalation exposure. Saf. Health Work 2: 34–38.

Kim, J.A., C. Åberg, A. Salvati and K.A. Dawson. 2012. Role of cell cycle on the cellular uptake and dilution of nanoparticles in a cell population. Nature Nanotechnol. 7: 62–68.

Koao, L.F., F.B. Dejene and H.C. Swart. 2014. Synthesis of PbS nanostructures by chemical bath deposition method. Int. J. Electrochem. Sci. 9: 1747–1757.

Kong, J., C. Zhou, A. Morpurgo, H.T. Soh, C.F. Quate and C. Marcus. 1999. Synthesis, integration, and electrical properties of individual single-walled carbon nanotubes. Appl. Phys. A Mater. 69: 305–308.

Kovacic, P. and R. Somanathan. 2010. Biomechanisms of nanoparticles (toxicants, antioxidants and therapeutics): Electron transfer and reactive oxygen species. J. Nanosci. Nanotechnol. 10: 7919–7930.

Kumbhakar, D.V., A.K. Datta, A. Mandal, D. Das, S. Gupta, B. Ghosh, S. Halder and S. Dey. 2016a. Effectivity of copper and cadmium sulphide nanoparticles in mitotic and meiotic cells of *Nigella sativa* L. (black cumin)–can nanoparticles act as mutagenic agents? J. Exp. Nano. Sci. 11: 823–839.

Kumbhakar, D.V., A.K. Datta, A. Mandal, D. Das, B. Ghosh, A. Pramanik and A. Saha. 2016b. Copper and cadmium sulphide nanoparticles can induce macromutation in *Nigella sativa* L. (black cumin) J. Plant Development Sci. 8: 371–377.

Kungumadevi, L. and R. Sathyamoorthy. 2013. Synthesis of PbTe nanocubes, worm-like structures and nanoparticles by simple thermal evaporation method. Bull. Mater. Sci. 36: 771–778.

Landsiedel, R., M.D. Kapp, M. Schulz, K. Wiench and F. Oesch. 2009. Genotoxicity investigations on nanomaterials: Methods, preparation and characterization of test material, potential artifacts and limitations–many questions, some answers. Mutat. Res. 681: 241–258.

Langie, S.A.S., A. Azqueta and A.R. Collins. 2015. The comet assay: Past, present, and future. Front. Genet. 6: 266.

Laruea, C., H. Castillo-Michelb, S. Sobanskac, N. Trcerad, S. Sorieule, L. Cécillona, L. Ouerdanef, S. Legrosg and G. Sarreta. 2014. Fate of pristine TiO_2 nanoparticles and aged paint-containing TiO_2 nanoparticles in lettuce crop after foliar exposure. J. Hazardous Mat. 273: 17–26.

Lee, J.Y., B.H. Hong, W.Y. Kim, S.K. Min, Y. Kim, M.V. Jouravlev, R. Bose, K.S. Kim, I-C. Hwang, L.J. Kaufman, C.W. Wong, P. Kim and K.S. Kim. 2009. Near-field focusing and magnification through self-assembled nanoscale spherical lenses. Nature 460: 498–501.

Lee, W., S.H. Kang, J-Y. Kim, G.B. Kolekar, Y-E. Sung and S-H. Han. 2009. TiO$_2$ nanotubes with a ZnO thin energy barrier for improved current efficiency of CdSe quantum-dot-sensitized solar cells. Nanotechnology 20: 335706–335713.

Lee, C.W., S. Mahendra, K. Zodrow, D. Li, Y.C. Tsai and J. Braam. 2010. Developmental phytotoxicity of metal oxide nanoparticles to *Arabidopsis thaliana*. Environ. Toxicol. Chem. 29: 669–675.

Li, Z. and D. Xing. 2011. Mechanistic study of mitochondria-dependent programmed cell death induced by aluminium phytotoxicity using fluorescence techniques. J. Exp. Bot. 62: 331–343.

Li, J., B. Arnal, C-W. Wei, J. Shang, T-M. Nguyen, M. O'Donnell and G. Xiaohu. 2015. Magneto optical nanoparticles for cyclic magnetomotive photoacoustic imaging. ACS Nano. 9: 1964−1976.

Lidén, G. 2011. The European commission tries to define nanomaterials. Ann. Occup. Hyg. 55: 1–5.

Limbach, L.K., Y. Li, R.N. Grass, T.J. Brunner, M.A. Hintermann, M. Muller, D. Gunther and W.J. Stark. 2005. Oxide nanoparticle uptake in human lung fibroblasts. Effect of particle size, agglomeration, and diffusion at low concentrations. Environ. Sci. Technol. 39: 9370–9376.

Lin, B.S., S.Q. Diao, C.H. Li, L.J. Fang, S.C. Qiao and M. Yu. 2004. Effect of TMS (nanostructured silicon dioxide) on growth of Changbai larch seedlings. J. For. Res. 15: 138–140.

Liu, J.B., H. Dong, Y.M. Li, P. Zhan, M.W. Zhu and Z.L. Wang. 2006. A facile route to synthesis of ordered arrays of metal nanoshells with a controllable morphology. Jpn. J. Appl. Phys. 45: L582–L584.

Liu, Q., Y. Zhao, Y. Wan, J. Zheng, X. Zhang, C. Wang, X. Fang and J. Lin. 2010. Study of the inhibitory effect of water-soluble fullerenes on plant growth at the cellular level. ACS Nano. 4: 5743–5748.

Liu, R.Q., H.Y. Zhang and R. Lal. 2016. Effects of stabilized nanoparticles of copper, zinc, manganese, and iron oxides in low concentrations on lettuce (*Lactuca sativa*) seed germination: Nanotoxicants or nanonutrients? Water Air Soil Pollut. 227: 1–14.

López-Moreno, M.L., G. de la Rosa, J.A. Hernández-Viezcas and J.R. Peralta-Videa. 2010. Gardea-Torresdey. X-ray Absorption Spectroscopy (XAS) Corroboration of the uptake and storage of CeO$_2$ nanoparticles and assessment of their differential toxicity in four edible plant species. J. Agric. Food Chem. 58: 3689–3693.

Lv, J., S. Zhang, L. Luo, J. Zhang, K. Yang and P. Christie. 2015. Accumulation, speciation and uptake pathway of ZnO nanoparticles in maize. Environ. Sci. Nano. 2: 68–77.

Ma, X., J.G. Lee, Y. Deng and A. Kolmakov. 2010. Interactions between engineered nanoparticles (ENPs) and plants: Phytotoxicity, uptake and accumulation. Sci. Total Environ. 408: 3053–3061.

Ma, Y.H., L.L. Kuang, H. Xiao, B. Wei, Y.D. Ya, Y.Z. Zhi, L.Z. Yu and F.C. Zhi. 2010. Effect of rare earth oxide nanoparticles on root elongation of plants. Chemosphere 78: 273−279.

Ma, C., H. Liu, H. Guo, C. Musante, S.H. Coskun, B.C. Nelson, J.C. White, B. Xing and O.P. Dhankher. 2016. Defence mechanisms and nutrient displacement in *Arabidopsis thaliana* upon exposure to CeO$_2$ and In$_2$O$_3$ nanoparticles. Environ. Sci. Nano. 3: 1369–1379.

Mahmoudi, M., K. Azadmanesh, M.A. Shokrgozar, W.S Journeay and S. Laurent. 2011. Effect of nanoparticles on the cell life cycle. Chem. Rev. 111: 3407–3432.

Majumdar, S., J.R. Peralta-Videa, S. Bandyopadhyay, H. Castillo-Michel, J.A. Hernandez-Viezcas, S. Sahi and J.L. Gardea-Torresdey. 2014. Exposure of cerium oxide nanoparticles to kidney bean shows disturbance in the plant defence mechanisms. J. Hazard. Mater. 278: 279–287.

Masarovičová, E. and K. Kráľová. 2013. Metal nanoparticles and plants. Ecol. Chem. Eng. S. 20: 9–22.

Meriga, B., B.K. Reddy, R.R. Kalashikam and P.B.K. Kishor. 2004. Aluminium-induced production of oxygen radicals, lipid peroxidation and DNA damage in seedlings of rice (*Oryza sativa*). J. Plant Physiol. 161: 63–68.

Mishra, J.K. and R. Adelung. 2017. ZnO tetrapod materials for functional applications. Mater. Today. https://doi.org/10.1016/j.mattod.2017.11.003.

Mittler, R. 2002. Oxidative stress, antioxidants and stress tolerance. Trends in Plant Science 7: 405–410.

Mokerov, V.G., Y.V. Fedorov, L.E. Velikovski and M.Y. Scherbakova. 2001. New quantum dot transistor. Nanotechnology 12: 552–555.

Mostofizadeh, A., Y. Li, B. Song and Y. Huang. 2011. Synthesis, properties, and applications of low-dimensional carbon-related nanomaterials. J. Nanomat. 2011: 1–21.

Nair, R., S.H. Varghese, B.G. Nair, T. Maekawa, Y. Yoshida and D. Sakthi Kumar. 2010. Nanoparticulate material delivery to plants. Plant Sci. 179: 154–163.

Nair, P.M.G. and I.M. Chung. 2015. Changes in the growth, redox status and expression of oxidative stress related genes in chickpea (*Cicer arietinum* L.) in response to copper oxide nanoparticle exposure. J. Plant Growth Regul. 34: 350–361.

Nel, A., T. Xia, L. Madler and N. Li. 2006. Toxic potential of materials at the nanolevel. Science 311: 622–627.

Nguyen, H.L., H.N. Nguyen, H.H. Nguyen, M.Q. Luu and M.H. Nguyen. 2015. Nanoparticles: Synthesis and applications in life science and environmental technology. Adv. Nat. Sci.: Nanosci. Nanotechnol. 6: 015008–015017.

Nowack, B. and T.D. Bucheli. 2007. Occurrence, behavior and effects of nanoparticles in the environment. Environ. Pollut. 150: 5–22.

Padmavathy, N. and R. Vijayaraghavan. 2008. Enhanced bioactivity of ZnO nanoparticles—an antimicrobial study. Sci. Technol. Adv. Mater. 9: 035004–035010.

Pakrashi, S., N. Jain, S. Dalai, J. Jayakumar, P.T. Chandrasekaran, A.M. Raichur, N. Chandrasekaran, A. Mukherjee, and V. Bansal. 2014. *In vivo* genotoxicity assessment of titanium dioxide nanoparticles by *Allium cepa* root tip assay at high exposure concentrations. PLoS One 9: e87789.

Pan, J.W., M.Y. Zhu and H. Chen. 2001. Aluminum-induced cell death in root-tip cells of barley. Environ. Exp. Bot. 46: 71–79.

Pan, M., S. Xing, T. Sun, W. Zhou, M. Sindoro, H.H. Teo, Q. Yan and H. Chen. 2010. 3D dendritic gold nanostructures: Seeded growth of a multi-generation fractal architecture. Chem. Commun. 46: 7112–7114.

Panda, K.K., D. Golari, A. Venugopal, V. Mohan, M. Achary, G. Phaomei, N.L. Parinandi, H.K. Sahu and B.B. Panda. 2017. Green Synthesized Zinc Oxide (ZnO) nanoparticles induce oxidative stress and DNA damage in *Lathyrus sativus* L. root bioassay system. Antioxidants 6: 35–51.

Papavlassopoulos, H., Y.K. Mishra, S. Kaps, I. Paulowicz, R. Abdelaziz, M. Elbahri, E. Maser, R. Adelung and C. Röhl. 2014. Toxicity of functional nano-micro zinc oxide tetrapods: Impact of cell culture conditions, cellular age and material properties. PLoS ONE 9: e84983.

Passardi, F., C. Cosio, C. Penel and C. Dunand. 2005. Peroxidases have more functions than a Swiss army knife. Plant Cell Rep. 24: 255–265.

Patlolla, A.K., A. Berry, L. May and P.B. Tchounwou. 2012. Genotoxicity of silver nanoparticles in *Vicia faba*: A pilot study on the environmental monitoring of nanoparticles. Int. J. Environ. Res. Public Health 9: 1649–1662.

Phugare, S.S., D.C. Kalyani, A.V. Patil and J.P. Jadhav. 2011. Textile dye degradation by bacterial consortium and subsequent toxicological analysis of dye and dye metabolites using cytotoxicity, genotoxicity and oxidative stress studies. J. Hazard. Mater. 186: 713–723.

Pourrut, B., E. Pinelli, V. Celiz Mendiola, J. Silvestre and F. Douay. 2015. Recommendations for increasing alkaline comet assay reliability in plants. Mutagenesis 30: 37–43.

Pradhan, D. and K.T. Leung. 2008. Vertical growth of two-dimensional zinc oxide nanostructures on ITO-coated glass: effects of deposition temperature and deposition time. J. Phys. Chem. 112: 1357–1364.

Pramanik, A., A.K. Datta, S. Gupta, B. Ghosh, D. Das and D.V. Kumbhakar. 2017. Assessment of genotoxicity of engineered nanoparticles (cadmium sulphide–CdS and copper oxide–CuO) using plant model (*Coriandrum sativum* L.). Int. J. Res. Pharm. Sci. 8: 1–13.

Pramanik, A., A.K. Datta, D, Das, D.V. Kumbhakar, B. Ghosh, A. Mandal, S. Gupta, A. Saha and S. Sengupta. 2018. Assessment of mutagenic potentiality of chemically synthesized cadmium sulphide and copper oxide nanoparticles in *Coriandrum sativum* L. (Apiaceae). Cytology and Genetics (in Press).

Prokhorova, I.M., B.S. Kibrik, A.V. Pavlov and D.S. Pesnya. 2013. Estimation of mutagenic effect and modifications of mitosis by silver nanoparticles. Bull. Exp. Biol. Med. 156: 255–259.

Rajeshwari, A., S. Kavitha, S.A. Alex, D. Kumar, A. Mukherjee, N. Chandrasekaran and A. Mukherjee. 2015. Cytotoxicity of aluminium oxide nanoparticles on *Allium cepa* root tip–effects of oxidative stress generation and biouptake. Environ. Sci. Pollut. Res. 22: 11057–11066.

Raliya, R., J.C. Tarafdar, K. Gulecha, K. Choudhary, R. Ram, P. Mal and R.P. Saran. 2013. Review article; scope of nanoscience and nanotechnology in agriculture. J. Appl. Biol. Biotechnol. 1: 41–44.

Raliya, R. and J.C. Tarafdar. 2013. ZnO nanoparticle biosynthesis and its effect on phosphorous-mobilizing enzyme secretion and gum contents in clusterbean (*Cyamopsis tetragonoloba* L.). Agric. Res. 2: 48–57.

Raliya, R., C. Franke, S. Chavalmane, R. Nair, N. Reed and P. Biswas. 2016. Quantitative understanding of nanoparticle uptake in watermelon plants. Front. Plant Sci. 7: 1288.

Rao, K., A. Raghavendra and K. Reddy. 2006. Physiology and molecular biology of stress tolerance in plants. Netherlands, Springer.

Rao, S. and G.S. Shekhawat. 2016. Phytotoxicity and oxidative stress perspective of two selected nanoparticles in *Brassica juncea*. 3 Biotech. 6: 244.

Raskar, S.V. and S.L. Laware. 2014. Effect of zinc oxide nanoparticles on cytology and seed germination in onion. Int. J. Curr. Microbiol. Appl. Sci. 3: 467–473.

Remédios, C., F. Rosário and V. Bastos. 2012. Environmental nanoparticles interactions with plants: morphological, physiological, and genotoxic aspects. J. Bot. 2012: 1–8.

Rico, C.M., S. Majumdar, M. Duarte-Gardea, J.R. Peralta-Videa and J.L. Gardea-Torresdey. 2011. Interaction of nanoparticles with edible plants and their possible implications in the food chain. J. Agric. Food Chem. 59: 3485–3498.

Roberts, A.G. and K.J. Oparka. 2003. Plasmodesmata and the control of symplastic transport. Plant Cell Environ. 26: 103–124.

Roco, M.C. 2003. Broader societal issues of nanotechnology. J. Nanopart. Res. 5: 181–189.

Rothen-Rutishauser, B.M., S. Schurch, B. Haenni, N. Kapp and P. Gehr. 2006. Interaction of fine particles and nanoparticles with red blood cells visualized with advanced microscopic techniques. Environ. Sci. Technol. 40: 4353–4359.

Ruso, J.M., V. Verdinelli, N. Hassan, O. Pieroni and P.V. Messina. 2013. Enhancing CaP biomimetic growth on TiO$_2$ cuboids nanoparticles via highly reactive facets. Langmuir. 29: 2350–2358.

Saha, N. and S. Dutta Gupta. 2017. Low-dose toxicity of biogenic silver nanoparticles fabricated by *Swertia chirata* on root tips and flower buds of *Allium cepa*. J. Hazard. Mater. 330: 18–28.

Scandalios, J.G. 1993. Oxygen stress and superoxide dismutases. Plant Physiol. 101: 7–12.

Schwab, F., G. Zhai, M. Kern, A. Turner, J.L. Schnoor and M.R. Wiesner. 2015. Barriers, pathways and processes for uptake, translocation and accumulation of nanomaterials in plants—critical review. Nanotoxicol. 10: 257–278.

Scrinis, G. and K. Lyons. 2007. The emerging nano-corporate paradigm: Nanotechnology and the transformation of nature, food and agri-food systems. Int. J. Sociol. Agric. Food 15: 22–44.

Selivanov, V.N. and E.V. Zorin. 2001. Sustained action of ultrafine metal powders on seeds of grain crops. Perspekt. Materialy 4: 66–69.

Sharma, P., A.B. Jha, R.S. Dubey and M. Pessarakli. 2012. Reactive oxygen species, oxidative damage, and antioxidative defence mechanism in plants under stressful conditions. J. Bot. 2012: 217037.

Shaymurat, T., J. Gu, C. Xu, Z. Yang, Q. Zhao, Y. Liu and Y. Liu. 2012. Phytotoxic and genotoxic effects of ZnO nanoparticles on garlic (*Allium sativum* L.): A morphological study. Nanotoxicol. 6: 241–248.

Siddiqi, K.S. and A. Husen. 2017. Plant response to engineered metal oxide nanoparticles. Nanoscale Res. Lett. 12: 92.

Singh, N., B. Manshian, J.S. Gareth, G.J.S. Jenkins, S.M. Griffiths, P.M. Williams, T.G.G. Maffeis, C.J. Wright and S.H. Doak. 2009. Nanogenotoxicology: The DNA damaging potential of engineered nanomaterials. Biomater. 30: 3891–3914.

Singh, S. 2013. Zinc oxide nanostructures: Synthesis, characterizations and device applications. J. Nanoeng. Nanomanuf. 3: 1–28.

Smirnoff, N. 2011. Vitamin C: The metabolism and functions of ascorbic acid in plants. Adv. Bot. Res. 59: 107–177.

Stouwdam, J.W. and R.A.J. Janssen. 2008. Red, green, and blue quantum dot LEDs with solution processable ZnO nanocrystal electron injection layers. J. Mater. Chem. 18: 1889–1894.

Sun, D., H.I. Hussain, Z. Yi, R. Siegele, T, Cresswell, L. Kong and D.M. Cahill. 2014. Uptake and cellular distribution, in four plant species, of fluorescently labelled mesoporous silica nanoparticles. Plant Cell Rep. 33: 1389–1402.

Tanou, G., A. Molassiotis and G. Diamantidis. 2009. Induction of reactive oxygen species and necrotic death-like destruction in strawberry leaves by salinity. Environ. Exp. Bot. 65: 270–281.

Tervonen, T., I. Linkov, J.R. Figueira, J. Steevens, M. Chappell and M. Merad. 2009. Risk-based classification system of nanomaterials. J. Nanopart. Res. 11: 757–766.

Thomas, J.P., M. Maiorinog, F. Ursinis and A.W. Girotti. 1990. Protective action of phospholipid hydroperoxide glutathione peroxidase against membrane-damaging lipid peroxidation. J. Biol. Chem. 265: 454–461.

Tiwari, J.N., R.N. Tiwari and K.S. Kim. 2012. Zero-dimensional, one-dimensional, two-dimensional and three-dimensional nanostructured materials for advanced electrochemical energy devices. Prog. Mater. Sci. 57: 724–803.

Tripathi, D.K., S. Singh, S. Singh, P.K. Srivastava, V.P. Singh, S. Singh, P.K. Singh, S.M. Prasad, A.C. Pandey, D.K. Chauhan and N.K. Dubey. 2017. Nitric oxide alleviates silver nanoparticles (AgNPs)-induced phytotoxicity in *Pisum sativum* seedlings. Plant Physiol. Biochem. 110: 167–177.

Tripathy, B.C. and R. Oelmüller. 2012. Reactive oxygen species generation and signalling in plants. Plant Signal. Behav. 7: 1621–1633.

Urrutia, A., J. Goicoechea and F.J. Arregui. 2015. Optical fiber sensors based on nanoparticle-embedded coatings. J. Sensors 2015: 1–18.

Ustinov, V.M., A.E. Zhukov, A.R. Kovsh, S.S. Mikhrin, N.A. Maleev, B.V. Volovik, Y.G. Musikhin, Y.M. Shernyakov, E.Y. Kondat'eva and M.V. Maximov. 2000. Long-wavelength quantum dot lasers on GaAs substrates. Nanotechnol. 11: 397–400.

Vamanu, C.I., M.R. Cimpan, P.J. Hol, S. Sornes, S.A. Lie and N.R. Gjerdet. 2008 Induction of cell death by TiO₂ nanoparticles: Studies on a human monoblastoid cell line. Toxicol. *In Vitro* 22: 1689–1696.

Van Dongen, J.T., A.M.H. Ammerlaan, M. Wouterlood, A.C.V. Van Aelst and A.C. Borstlap. 2003. Structure of the developing pea seed coat and the post-phloem transport pathway of nutrients. Ann. Bot. 91: 729–737.

Vecchio, G., A. Galeone, V. Brunetti, G. Maiorano, L. Rizzello, S. Sabella, R. Cingolani and P.P. Pompa. 2011. Mutagenic effects of gold nanoparticles induce aberrant phenotypes in *Drosophila melanogaster*. Nanomedicine 8: 1–7.

Vidal, A.E., I.D. Hickson, I. Boiteux and J.P. Radicella. 2001. Mechanism of stimulation of the DNA glycosylase activity of hOGG1 by the major human AP endonuclease: Bypass of the AP lyase activity step. Nucleic Acids Res. 29: 1285–1292.

Wang, J., M. Lin, Y. Yan, Z. Wang, P.C. Ho and K.P. Loh. 2009. CdSe/AsS core-shell quantum dots: Preparation and two-photon fluorescence. J. Am. Chem. Soc. 131: 11300–11301.

Wang, H., H.S. Casalonque, Y. Liang and H. Dai. 2010. Ni(OH)₂ nanoplates grown on graphene as advanced electrochemical pseudocapacitor. Materials. J. Am. Chem. Soc. 132: 7472–7479.

Wu, W., Q. He and C. Jiang. 2008. Magnetic iron oxide nanoparticles: Synthesis and surface functionalization strategies. Nanoscale Res. Lett. 3: 397–415.

Xi, Z.G., F.H. Chao, D.F. Yang, Y.M. Sun, G.X. Li, H.S. Zhang, W. Zhang, Y.H. Yang and H.L. Liu. 2004. The effects of DNA damage induced by acetaldehyde. Huan Jing Ke Xue 25: 102–105.

Xu, S. and Z.L. Wang. 2011. One-Dimensional ZnO nanostructures: Solution growth and functional properties. Nano Res. 4: 1013–1098.

Yang, J., W. Cao and Y. Rui. 2017. Interactions between nanoparticles and plants: Phytotoxicity and defence mechanisms. J. Plant Interact. 12: 158–169.

Zagorchev, L., C.E. Seal, I. Kranner and M. Odjakova. 2013. A central role for thiols in plant tolerance to abiotic stress. Int. J. Mol. Sci. 14: 7405–7432.

Zhang, G. and D. Wang. 2008. Fabrication of heterogeneous binary arrays of nanoparticles via colloidal lithography. J. Am. Chem. Soc. 130: 5616–5617.

Zhang, P., Y. Ma, S. Liu, G. Wang, J. Zhang, X. He, J. Zhang, Y. Rui and Z. Zhang. 2017. Phytotoxicity, uptake and transformation of nano-CeO$_2$ in sand cultured romaine lettuce. Environ. Pollut. 220: 1400–1408.

Zheludkevicha, M.L., R. Serraa, M.F. Montemorb and M.G.S. Ferreiraab. 2005. Oxide nanoparticle reservoirs for storage and prolonged release of the corrosion inhibitors. Electrochem. Commun. 7: 836–840.

Zhu, H., J. Han, J.Q. Xiao and Y. Jin. 2008. Uptake, translocation, and accumulation of manufactured iron oxide by pumpkin plants. J. Environ. Monit. 10: 713–717.

Index

Printed and bound by CPI Group (UK) Ltd, Croydon, CR0 4YY

24/10/2024

01778307-0001